TEXTBOOK OF ENVIRONMENTAL ENGINEERING

P. VENUGOPALA RAO
Principal
Sridevi Women's Engineering College
Hyderabad

PHI Learning Private Limited
Delhi-110092
2024

*In fond memory of **Shri Asoke K. Ghosh** (October 1942 – February 2024), Founder Chairman and Managing Director of PHI Learning, whose vision endlessly inspires.*

The Legacy Continues. . . .

Published by Pushpita Ghosh, PHI Learning Private Limited, Rimjhim House, 111, Patparganj Industrial Estate, Delhi-110092 and Printed by Multi Colour Services, I-45, DLF Industrial Area, Sector-32, Faridabad, Haryana-121003.

₹625.00

TEXTBOOK OF ENVIRONMENTAL ENGINEERING
by P. Venugopala Rao

ISBN-978-81-203-1930-1 (Print Book)
ISBN-978-93-90669-24-0 (e-Book)

To my Parents

Sri P. Sambasiva Rao** and **Smt. P. Sitaramamma

*and my wife **Smt. P. Suseela***

Contents

Preface

The living conditions today have surely improved to a great deal due to industrialization and urbanization, but only at the cost of environmental degradation. The resources are not only getting depleted, but are also getting polluted; the wastes generated are mostly hazardous and the air is getting filled with pollutants due to high consumption of fossil fuels. With the growing concerns about the environment, the newer environmental issues regarding hazardous waste, global climate change, stratospheric ozone depletion, ground water contamination, and disaster mitigation have become the focus of much of the environmental attention in more recent years. As the very nature of the problems has changed, scientists and engineers are looking at them from different angles. For example, when we find potential carcinogens in our ground water at many places, we are forced to realize that we ourselves have been creating toxic chemicals and carcinogens by various chemical processes such as chlorinating our surface water, emitting CFCs in the atmosphere, and so forth. Therefore, attempts are being made by the concerned authorities to create a greater awareness among the present generation by disseminating the knowledge and information on how to protect our environment through formalizing various courses for the management personnel as well as engineering and science students. Although there are numerous scientific journals and textbooks available for advanced students, most of these materials do not include the fundamental engineering principles that a beginning student probably would want to learn. This comprehensive text bridges the gap between the general introductory topics and the more advanced environmental engineering issues.

The book is an outgrowth of the author's rich and varied experience in teaching the subject at various levels (B. Tech, M. Tech, M.Sc.) in Jawaharlal Nehru Technological University (JNTU), Hyderabad, for over 35 years.

Designed for a two-semester course on the essential environmental engineering aspects for undergraduate engineering and postgraduate science students, the text will prove to be highly useful for managers, administrators and environmentalists who are engaged in finding the control methodologies of air and water pollution. Those who are teaching this subject may also find the book quite helpful.

With a student-oriented approach, the book is organized into eight chapters and three appendices. Chapters 1 and 2 respectively contain the two basic aspects of environment—Ecology and Environmental Sanitation. Ecology includes both living and non-living entities in relation to their surroundings, besides topics such as the biogeochemical cycles in nature and energy flow. Environmental sanitation, a topic that is not generally covered in most textbooks, discusses issues such as occupational hazards, noise control, soil pollution and refuse disposal with a focus on community health and principles of sanitation. Chapter 3 (Drinking Water) and Chapter 4 (Domestic Sewage) along with the two initial chapters are mostly taught in the first semester. Topics like collection, purification and distribution of drinking water may be covered in a 15–20 lecture schedule. Sufficient care has been taken to present only the relevant matter eliminating the not-so-important issues. The key principles of engineering that are required for the treatment and supply of water have been illustrated with a number of worked-out examples. In Chapter 4, design and construction of sewerage system have been dealt with in detail apart from sewage characteristics and its treatment. Waste water treatment, which is the backbone of environmental engineering, has been given a thorough look with emphasis on design requirements and fundamental principles to the extent that is required at this level.

Chapter 5, dealing with treatment of industrial wastes, is in a way an extension of Chapter 4. So the required information has been presented in a consolidated form. As this subject is industry specific, it requires a further detailed study. Chapter 6 on Air Pollution includes various topics that are systematically arranged with neat illustrations. The methods of monitoring and analysis of air quality as well as water and waste water have been discussed in Chapter 7, giving the relevant laboratory procedure for the analysis. In Chapter 8, the fundamental concepts of environmental impact analysis (EIA) have been specially introduced to help students find the solutions to the environmental problems. The collection of data from various industries which adopt the processes to recover/reuse material from the industrial effluents will be quite useful. Environmental audit (EA) and functions of the Pollution Control Board will familiarize the reader with the audit procedure and environmental management planning. The topics of these chapters are mostly taken up in the second semester. Toxic wastes, application of geographical information system and disaster management—the three burning issues—have been briefly discussed in the appendices.

As the topics in the book are self-explanatory, students need not therefore refer to a number of other books on the subject. Each chapter gives a summary of the key points discussed in the text and concludes with objective type questions enabling the students to measure their comprehension.

The author expresses his gratitude to Prof. C.A. Sastry (IIT Madras), Prof. A.V.S. Prabhakar Rao (IIT Kanpur), Prof. G. Ram Prasad (SVUCE, Tirupathi), Prof. Y. Anjaneyulu (JNTU, Hyderabad) and Smt. A. Vijayalakshmi (BMS Ec, Bangalore) who have shown their concern by offering valuable suggestions. Thanks are also due to Shri Ashok Rao and his family members with whom the author is presently associated with. Finally, the author conveys his sincere thanks to the publishers, Prentice-Hall of India whose encouragement and assistance have helped in the speedy and quality publication of the book.

P. VENUGOPALA RAO

CHAPTER 1

Ecology

1.1 INTRODUCTION

1.1.1 Biosphere

The living environment, known as *biosphere*, which consists of land (lithosphere), water (hydrosphere), and air (atmosphere), is the basis for the sustenance and continuance of life. Plants, animals, and all organisms live together, interacting among themselves and with the environment.

Food and energy, which are essential for life, are obtained from the surrounding environmental resources. Local environmental conditions depend on a variety of factors such as the nature of soil, rainfall, temperature, wind pressure, land and water bodies, distance from the sea, altitude of the place, vegetation and length of the day. Weather and climatic conditions at a place are decided mostly by geographical location (latitude and longitude) and topography. Environmental conditions and the utilisation of environmental resources are not the same in various parts of the world. As living conditions vary, different types of habitats exist.

Land

Soil, the outer crust of the earth, contains minerals, organic matter, water and air, which are essential for the growth and development of plants, animals and human beings. Soil is the medium for nutrients and detritus food chain. The soil-water-air relationship is responsible for the plant productivity. Ca, Mg, Na, K, Fe, Al, Zn, C, H, O, N, P, S and many trace elements are plant nutrients. Fertilizers (N, P, K) are absorbed through soil. Fertilizers and pesticides used in raising crops reach water resources through land.

Water

While plenty of water is available in rivers, lakes, reservoirs, oceans, only about 1 per cent of fresh water is available for human use in the shape of surface and ground water. Water is used for irrigation, power production, industry, transport, in addition to domestic use. Water sources are in contact with soil and dissolve many salts and nutrient minerals depending on their solubility. Water supports also fish and other aquatic life.

Air

Air contains nitrogen (78 per cent), oxygen (21 per cent) and traces of a number of other gases. It is always under circulation in the troposphere (about 10–15 km from the top of land surface).

Earth absorbs radiation in the visible region and emits radiation back in the infra-red region. Temperature on the surface of earth is maintained approximately constant, due to reabsorption of the outgoing radiation by water vapour, carbon dioxide, and scattering of sunlight into atmospheric space. Oxygen level is maintained as carbon dioxide is utilised by plants during photosynthesis. Ozone layer provides a protective blanket on earth.

1.1.2 Scope

Ecology deals with living and non-living entities in relation to their surroundings, in the total environment. Men, animals, birds, plants, fish, insects, microorganisms, fungi, algae, etc. form the living environment. Their co-existence is due to interdependency.

Soil, water, air and nutrients, constitute the non-living component. They are the essential resources in the living environment. Sun is the chief source of energy in the complete system.

Notation

Organisms. Any living entity, animal or plant, capable of growth and reproduction, is known as *organism*. These organisms directly interact with each other and with the surrounding environment.

Species. A group of individual organisms having common attributes (characteristics not shown by others) are termed as *species*.

Populations. These are groups of genetically similar individuals utilising the same resources in a similar fashion.

Communities. Groups of populations co-occurring in space and time are referred to as *communities*. They form a colony for carrying out different activities.

Habitat. It essentially means a natural home, a place where environmental factors are present and natural resources are evenly utilised. Different kinds of organisms live together for mutual benefit.

Ecosystem. A group of organisms interacting among themselves and with the surrounding environment forms an *ecosystem* which includes both livng and non-living components of the environment. Bio-diversity is a beauty of the surroundings and an environmental necessity.

Environment normally includes the physical, chemical, biological, cultural, economic, social and other characteristics of a location. Environmental resources like water, soil, air, nutrients, minerals, etc. are renewable to a certain extent, but the carrying capacity of the environment (the capacity to support maximum number of living organisms in a given time and space) is not infinite. As long as the utilisation of resources does not exceed the rate of regeneration, the environment is said to be 'sustainable'. This implies that populations always need a clean and fresh environment for their safe living. Poor quality (or depletion) of resources affects the survival and growth characteristics of living communities.

1.1.3 Ecosystems

Ecosystems are unique living locations. They are the basic structural and functional units of nature. Several communities exist in a particular area, under similar climatic and environmental conditions. Each group of organisms has a specific living space (*niche*) and interacts with other groups. The inputs into an ecosystem are water, air, soil, and solar energy. Ecosystems are not independent, separate entities, but are interrelated. Energy and nutrients flow from one system to another. They have a unique quality of regeneration and self-purification for maintaining equilibrium conditions in the total system under natural conditions. Ecosystem structure and function depend on

(a) composition of biological communities (e.g. plants, animals, microbes)

(b) quantity and distribution of materials such as nutrients, water, and

(c) existing range of living conditions, e.g. soil moisture, air temperature, photo period.

Life always utilises and depends on the non-living component in the environment. Particular biotic semblage exists for a given abiotic condition. Specific relationship exists between organisms of same or different kinds and environment. Ecosystems have diverse living conditions. Terrestrial system is different from aquatic life.

(a) Terrestrial biomes

Growth of vegetation depends on soil-water-atmospheric relationship. Regions like desert, tundra, grass land, savanna, coniferous and deciduous forests, tropical and rain forests are widely distributed throughout the globe.

Low rainfall and high evaporation cause desert climate, characterised by its shrubs, insects, birds and reptiles. Moderate and variable rainfall create grass lands, while good rainfall leads to development of tropical forests along equator and coniferous forests at high altitudes and latitudes. Species diversity is more prominent in the tropical forests. Types of vegetation, birds, insects and other organisms depend on the soil, rainfall, temperature, topography, geographical features, and other typical environmental conditions—the basis on which these biomes are formed. These conditions can transform a forest into a desert or vice versa. Vegetation (biomass) supplies nutrients and energy in these biomes.

(b) Fresh water biomes

Movement and utilisation of nutrients vary with water depths in water bodies. Aquatic life depends on dissolved salts, oxygen availability in different layers, light intensity, thermal stratification, and other flow considerations.

Nutrient levels are different in rivers and lakes. Water is in continuous flow and mixes well in running water (lotic) biomes. More deposition and eutrophication take place in lake (lentic) waters. Thus, development of fish, diatoms, algae and other microorganisms varies, depending on zones of fresh water sections, lengths and depths and at various intervals of time.

(c) Marine biomes

Horizontal and vertical sections are categorised as distinct zones of sea. Different ecosystems exist in estuaries (coral reefs), near the shore line (littoral), near top water layers (neritic), and in the vast expanse of open sea (pelagic). Aquatic life varies depending on organic nutrients and organic detritus available in the ocean in different zones.

1.2 POPULATION REGULATION

Survival, growth, development and reproduction are the key characteristics of living beings. The whole process depends on the following:

1. Availability and suitability of food and nutrients
2. The nature and quality of soil, water and air
3. Factors such as climate and temperature, and
4. Presence of other organisms

1.2.1 Growth of Organisms

Increase in population at any location under normal environmental conditions, may follow a pattern which may be viewed in three ways:

1. Population grows upto a certain limit, i.e. the saturation point with no further increase, as shown in Fig. 1.1(a). It is similar to the increase in the process rate of photosynthesis depending upon the intensity of light.
2. Populations tend to increase first and decrease later in course of time, presenting an optimal trend based on the environmental conditions, as illustrated in Fig. 1.1(b). This is similar to the increase in the process rate of photosynthesis with increase in temperature.
3. Initially, populations may increase at a slow rate. Then, favourable conditions speed up the growth rate. In due course, an equilibrium stage (no further increase) is reached due to limitations of carrying capacity of

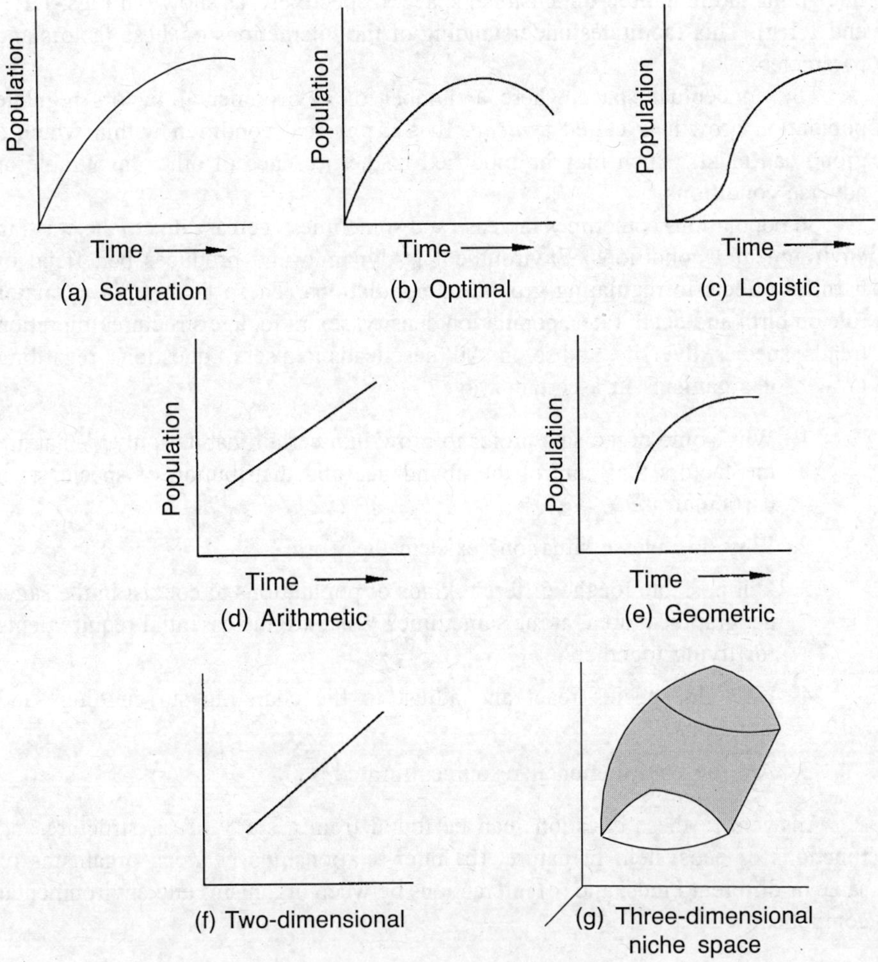

Fig. 1.1 Population regulation.

the environment (with regard to supply of energy and nutrients). Predator-prey relationship falls under this category of logistic (sigmoid) pattern as depicted in Fig. 1.1(c).

Simplifying the above facts, population growth may be considered to follow either a pattern similar to arithmetical progression as shown in Fig. 1.1(d) or geometrical progression as depicted in Fig. 1.1(e) depending on the availability of environmental resources.

A number of factors affect the population growth (as explained in sections 1.2.2 and 1.2.3). These are complex in nature, interactive and not mutually exclusive. The effect of one parameter may depend on the presence of other features. In fact, the actual growth or development is multifactorial with too many conflicts and overlaps.

The relationship between two factors and three parameters can be represented in a graph and in a three-dimensional space, respectively, as shown in Figs. 1.1(f) and 1.1(g). This facilitates understanding of the interactions of these factors and parameters.

The conceptual space where a number of environmental factors regulate population growth is called a *niche*. It is a physical condition within which a group can exist, which may be modified in the presence of other favourable or adverse conditions.

So populations sometimes increase and sometimes decrease due to the existing environmental conditions. Environment is dynamic and produces beneficial or harmful effects in regulating growth of populations. These factors play a major role on birth and death rates, population density, sex ratio, age structure, migration trends, species diversity, and so on. All these leads to a set of questions, regarding growth of organisms in a certain way:

1. Why some organisms prefer to grow in a certain habitat only? What are the factors that control the abundance and distribution of species in a particular area?
2. Why do diverse situations exist in the world?
3. Is it possible for the different kinds of populations to coexist in the same geographical area, at the same time? What are the essential requirements for living together?
4. How do species react and adjust to the changing surroundings and situations?
5. Are the environmental resources limited?

Answers to these questions can be found from a study of (a) structure and functions of ecosystem in nature, (b) inter-relationships between organisms of same or different kinds, and (c) interactions between organisms and environmental component.

1.2.2 Relationships

The following observations are useful for answering the above questions:

(i) *Co-existence.* Different species cannot co-exist for a long period if they depend on the same resource. They use the environment in different ways, by virtue of differences in their ecological niche. This segregation reduces competition and permits co-existence. Each ecosystem has a specific niche and interacts with others.

(ii) *Competition.* When resources deplete, species compete. Competition pressures determine the existence of a particular species in a community. Complete competitors do not co-exist; competition exists as interference, exploitation and infighting for the same food or need. In a struggle for existence, the fittest survives.

(iii) *Adjustment.* Organisms adjust to prevailing environmental conditions and limiting factors. Fish adapt and live, depending on the nature and quality of water available. They also adjust and acclimatise to all seasonal variations in temperature. Species migrate from one location to another searching for a suitable living environment.

(iv) *Development.* Algae, diatoms and bacteria live in pond medium. Light penetrates through shallow depths, due to sediments in water bodies. When less quantity of water is available, marshy vegetation develops, while woodlands are created if the soil is more fertile, and when rainfall and climate are very favourable, forests develop at those places. Similarly, animal life develops as fly, crustacean, fish to scorpion, duck, through frog, snake, etc., finally leading to a number of terrestrial animals. In the process of evolution, some species die, some may change, and a number of new genes may also come into existence. Succession is a directional change in community composition of ecosystems.

Volcanic ash formation, forest fires, earthquakes, nuclear blasts or such happenings may put to an end all theories related to the population growth.

1.2.3 Interactions

Interactions among various organisms (individuals of same or different species) are listed now. They affect the rate of growth of populations, in any of the following ways:

1. Beneficial to both groups living together (symbiosis, mutualism)
2. Harmful to others (antagonism)
3. Harming others without any benefit for it (amensalism)
4. Negative effect on each other (competition)
5. Big one eating small one (cannibalism)
6. Feeding on host organism (parasitism)
7. One is food for another (predation)
8. Producing something to destroy bacteria (antibiosis)

Association or segregation due to the above interactions depends on the individual community characteristics and neighbourhood relationship.

To sum up, population groups grow according to their own tolerance limits, and interactions with other organisms, based on the existing environmental conditions.

1.3 NATURAL CYCLES

A balancing action exists in the biosphere. Major biogeochemical cycles in nature are explained briefly in the following sections.

1.3.1 Hydrological Cycle

Rain water moves as surface runoff and ground water. Evaporation from water bodies, and transpiration from plant life result in formation of clouds, for further precipitation, as in the case of rain on earth's surface. Solar flux absorbed by earth's surface is the driving force for the hydrological (water) cycle as shown in Fig. 1.2.

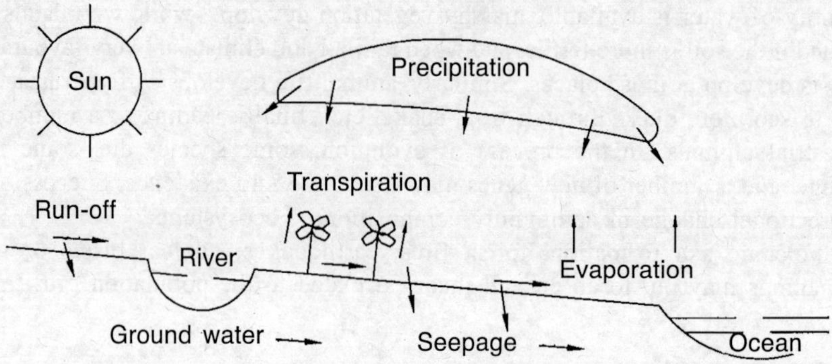

Fig. 1.2 Water cycle.

Water is a medium for biological activity. It is an agent for nutrient (chemicals and dissolved salts) distribution. Through precipitation and evaporation, water cycle acts as an agent for energy transfer and use. A desert or a forest formation is based on the availability of water at a place.

1.3.2 Carbon Cycle

Carbon is immobilised in the earth's crust (fossil fuel, carbonates). During photosynthesis carbon dioxide and nutrients are absorbed by plants (see section on Food Chain). Chemical combustion and respiration return the carbon to atmosphere. Decomposition of organic matter is another process in the carbon cycle. Ocean is a major sink for CO_2 in addition to the forest trees. Carbon cycle is presented in Fig. 1.3.

1.3.3 Nitrogen Cycle

Atmosphere, ocean, earth's crust and sediments are the reservoirs of nitrogen.

Nitrogen is to be fixed by microorganisms (Rhizobium) for utilisation by plants and animals. Significant biological involvement exists in both nitrification and denitrification processes.

Ammonification is the process of fixation of atmospheric nitrogen into NH_2 and NH_3 by bacteria. The reaction releases energy which is utilised by bacteria for their life process.

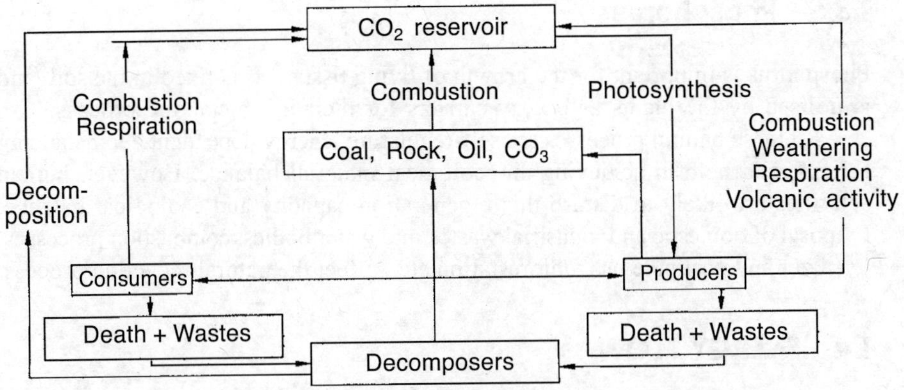

Fig. 1.3 Carbon cycle

Nitrification is the process of oxidising ammonia compounds to NO_2 and NO_3 by nitrosomonas and nitro-bacteria. If nitrates are not assimilated by plants, they remain in the soil and water sources.

Denitrification is the formation of N_2O, N_2 from nitrates (or NH_3 from NO_3) by Pseudomonas and other bacteria. Finally, molecular nitrogen splits into free nitrogen to join the atmosphere as shown in Fig. 1.4.

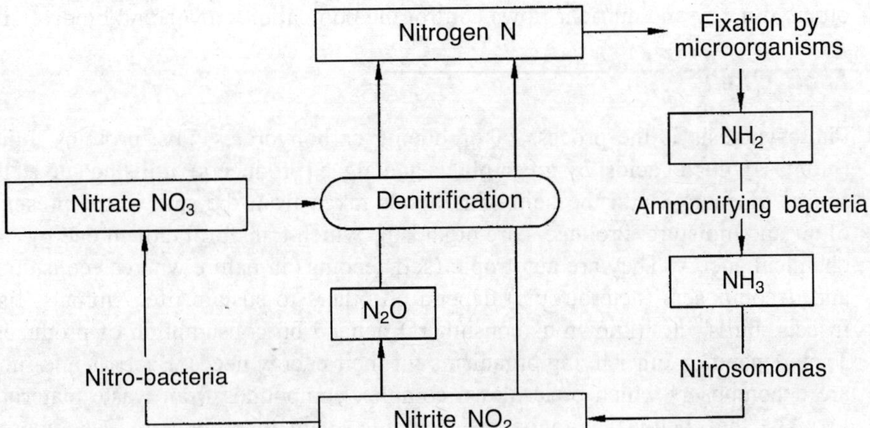

Fig. 1.4 Nitrogen cycle.

1.3.4 Sulphur Cycle

Atmosphere receives sulphur as H_2S (on reduction of sulphates), SO_2 (due to fossil fuel burning), and sulphates (due to ocean winds and volcanic emissions). Formation of sulphates is an aerobic action, and formation of H_2S is due to anaerobic decomposition.

1.3.5 Phosphorus

Phosphorus is important for the growth of living tissues. It is fixed in the soil, and is utilised by the plants as energy carriers for their life metabolic process.

All these natural processes are self-regulating, energy dependent and balancing actions, and help in achieving an ecological material balance. However, human activities are likely to disturb this regeneration capacity and ecological balance. Disposal of domestic and industrial wastes into water bodies, combustion processes, fertilizer and pesticide use, deforestation etc., affect the natural recycling process.

1.4 ENERGY FLOW

Population growth is regulated by the availability of food (nutrients) and stress tolerance limitations. Environment influences the life of organisms (e.g. photo periodism), and organisms modify the environment through their actions (e.g. nitrogen fixation).

Food and energy needs vary with species and situations. Organisms prosper in certain (niche) environmental conditions and the manifestation of energy and nutrient flows are the much needed ecological attributes.

Food chain is characterisation of energy and chemical flow through organisms. Nutrients and chemicals are returned to the environment and are available for others. Energy and nutrient flows control the population growth and biodiversity.

Food chain

Photosynthesis is the process of producing carbohydrates, fats, proteins, lipids (biomass), nuclei acids, by green plants and algae (producers), utilising sun as the source of energy with the help of minerals taken from the soil, in the presence of air and moisture. Producers are organisms which transform radiant energy into chemical energy. They are autotropic (self-feeding) in nature whereas consumers and decomposers (heterotropic) depend on others to sustain life. Animals, fish, insects, birds, etc. (known as consumers) depend on consumption of producers. They depend on other living organisms for their energy needs. Bacteria and fungi are decomposers which break down complex compounds from waste materials, to make the chemical components available again to producers. Decomposers utilise dead organisms. Primary production is based on soil fertility, water availability, temperature, day-light period, nutrients, oxygen, and other factors. Secondary production depends on the energy produced by the producers in the ecosystem.

Energy

Energy is stored in a chemical bond. Movement of energy and chemicals among tropic levels and between biotic and abiotic components integrates the ecosystem into a functional unit. Energy is the carrying capacity of the environment. Energy flow is always unidirectional (non-cyclic) along food chains. Ecosystem collapses if there is no continuous source of energy. Thus a dynamic ecosystem consists of energy flow, mineral cycling, interaction between physico-chemical environment, and biotic communities. Energy diminishes in the form of heat (respiration and excretion), or lost into other ecosystems through wind, water, mineral and animal movements. Energy is maintained through temperature regulation, interaction with organism, behavioural attitudes, replenishment of lost tissues etc. Steady state condition appears when inflows balance outflows of energy and nutrients in the system. In any system, energy can neither be created nor destroyed but it may change from one form to another. The sum of the total energy is constant. As energy moves in the system, it ultimately changes to heat, which is of no direct use to the system. Thus it is possible for dissipation of energy and the system losing its organised structure. Any system always moves towards a maximum disorder in the organisation. If input energy is added on a continuous basis, this entropy may decrease i.e. the system maintains an order in the net production. Environment quality parameters are operationally significant for the sustenance of ecosystems. Life on earth is linked directly to the quality of environment.

1.5 FORESTS AND WILD LIFE

Food, fodder, fuel, fibre, timber, etc., from the forests are an economic asset. Plants retain water. Trees provide protection from winds and prevent soil erosion. They absorb some pollutants including green house gases. CO_2-oxygen balance is maintained in the atmosphere due to photosynthesis. Trees reduce the impact of rainfall and runoff on soil. Forests also create a living environment for a variety of birds and animals. They are desirable for maintaining ecological stability and environmental equilibrium.

Deforestation leads to soil erosion, floods, droughts, less rainfall and desertification of lands. Commercial mining and quarrying, grazing, construction of roads, and reservoir projects are responsible for the loss of forest lands. More CO_2 in the atmosphere alters the percentage of reflected solar radiation of earth and leads to higher atmospheric temperatures.

1.6 HUMAN ACTIVITIES

Population growth, urbanisation and industrialisation disturb the ecological balance. More utilisation of resources may lead to depletion of natural resources. Ecosystem cannot survive without an effective nutrient supply and energy flow.

There are many human activities that affects nature. Some of these are as follows:

1. Dam (reservoir) construction leads to the submergence of valuable agricultural and forest lands, which leads to the loss of flora and fauna. Water logging and salinity are very important side effects.
2. Various industries discharge solid, liquid and gaseous effluents, polluting the land, water and atmospheric bodies. Ecosystems cannot survive when the limited carrying capacity of the environment is exhausted.

3. Pollutant concentration increases when they pass through water, soil, plants and aquatic life before they reach human beings (known as bio-magnification). Selenium poisoning (through soil plant) and mercury poisoning (through water fish) are well known.
4. Soil fertility is lost due to increased use of chemical fertilisers and pesticides.
5. Pollution from automobiles increases the level of CO and other pollutants in the ambient atmosphere.
6. Increased CO_2 levels in the atmosphere result in a rise in global temperatures.
7. More CFC in the atmosphere causes depletion of ozone layer as chlorine reacts with ozone.
8. Disposal of heavy metals and toxic chemicals leads to more pollution.
9. Radioactivity adversely affects the environment.

Therefore, pollution is an unfavourable alteration and causes a threat to life supporting systems. It disturbs the ecological balance.

Self-purification

Ecosystems are natural balancing systems over a period of time and space. This has been clearly depicted in the bio-geochemical cycles of nature. Waste is also an energy resource. Concentration of impurities in water or air is reduced due to mixing, dilution, dispersion and settling. Oxygen is utilised for the decomposition of organic matter in water bodies and again replenished from the atmosphere. Land acts as a filter for the removal of suspended solids. However, the capacity for self-purification (regeneration) is limited. Nature also maintains alternate renewable energy systems along with non-renewable energy sources.

Transportation or continuous movement in the biosphere helps in maintaining a logical self-sustaining capacity of the environment to a very reasonable extent. This carrying capacity of the environment may be utilised carefully, for discharging pollutants without stretching the limits of regeneration or recharging abilities of natural ecosystems.

1.7 CONCLUSION

There exists a specific relationship between plants and animals in the environment. Flow of nutrients and energy takes place as long as that specific relationship exists in the environment.

Forests are converted to grasslands and to desert land when the amount of rainfall reduces further. Aquatic life differs from fresh water ecosystems to marine waters (through estuaries) because of the salt content. Similarly, ponds and rivers (of different depths and reaches) have different life supporting systems. Change in the environment leads to changes in plant and animal lives; ultimately some may perish. The sustainability of environment depends on biodiversity and interdependency in the living environment. When these are affected, ecosystem, as a whole, slowly collapses. If the land is lost due to pollution, or ground water is contaminated, life in any form is endangered in that local area once the pollution level becomes higher than the tolerance levels of species. Aqua culture units, silting of lakes, mechanised slaughter houses, disposal of wastes cause considerable damage to the environment.

Man changes the environment through his actions. Plants, animals and materials are affected. Ultimately the environment controls the health and well-being of mankind. So there must be a symbiotic relationship between man and his environment for a sustainable living.

The following steps may safeguard the ecosystems:

1. Resource management in correlationship with the limitations of renewability.
2. Pollution control before disposal of waste into the environment.
3. Better land use patterns to protect the natural habitat.
4. Effective utilisation of the self-regulating (recycling) capacity of the environmental system.
5. Conserving the environmental assets for future generations.

It may be observed that progress and development should not be at the cost of polluting or depleting the existing environmental resources.

SUMMARY

- Environmental conditions are decided by the nature of soil, amount of rainfall, photoperiod and atmospheric temperature.
- Living and non-living things co-exist in an ecosystem (because of inter-dependency) in a specific manner.
- Population regulation and interactions among various organisms are too complex. However, a balancing force exists in nature.
- Pollution alters the ecosystem balance, affecting the nutrients and energy flow.
- Human activities should not disturb the biodiversity of an ecosystem and depletion of natural resources.
- Environmental assets need to be preserved for the sustenance and continuance of life.

REVIEW QUESTIONS

1. What are the essential requirements of an ecosystem?
2. How is ecological balance disturbed due to human activities?

3. Write a note on the factors affecting population growth.
4. Write short notes on
   (a) Nitrogen cycle
   (b) Sustainability and ecobalance
   (c) Interactions among various organisms

OBJECTIVE TYPE QUESTIONS

A. ***Each of the following questions comprises four alternatives. Select the most appropriate answer for each.***

1. Ecosystems
   (a) are independent units
   (b) have separate nutrient supplies
   (c) have self-purification system
   (d) have one group of organisms
2. Population growth is predicted by
   (a) gradual increasing trends upto a certain level
   (b) exponential increase in a given time period
   (c) logistic pattern
   (d) all the above
3. Which of the following produces negative effect on each other?
   (a) Antagonism
   (b) Parasitism
   (c) Competition
   (d) Predation

B. ***Fill in the blanks with suitable words:***

1. Living and non-living entities co-exist in the environment because _______.
2. Organisms live and interact in a specific living space known as _______.
3. Species diversity is more prominent in _______ type of forests.
4. In terrestrial biomes, _______ supplies nutrients.
5. Diverse situations exist in the world because _______.
6. A sink for CO_2 is _______.
7. Nitrogen fixation means _______.

CHAPTER 2

Environmental Sanitation

2.1 COMMUNITY HEALTH

2.1.1 Significance

Health does not mean mere absence of disease. It is defined as a state of complete physical, mental and social well being of an individual. When individuals are happy and healthy, the community prospers and the society as a whole becomes progressive. The living conditions and the working environment should therefore be ideal for the growth of the society and the nation.

Even as we make progress in some areas, we encounter new environmental challenges such as climate change, contaminated ground water, hazardous wastes, and degradation of air quality that affect people and ecosystems throughout the world. Water, air and land are the three main resources in our environment. To understand the nature of pollution problems, we must know the related physico-chemical reactions and only then can we find the engineering approaches to their solution. Solid, liquid and gaseous wastes are generated due to domestic, agricultural and industrial activities. These wastes are indiscriminately disposed on land, into water bodies and in ambient air. Atmosphere gets polluted with harmful chemicals and disease causing organisms. Diseases like cholera and typhoid are communicated from one person to another through water sources. Insects like flies and mosquitoes are also responsible for spreading diseases in the surrounding environment. As different groups of people meet at a festival, camp, school or in other public places, diseases are easily transmitted from the sick to the healthy individuals.

Prevention reduces the chances of getting infection of many common diseases, thus saving a person from the troubles of seeking medical treatment and suffering with the ill effects of the disease. So, adequate care has to be taken towards supplying pure water for domestic purposes and proper collection and disposal of solid wastes and sewage from the residential areas. Sanitation measures are mainly intended to control the various factors that are responsible for disease transmission and to provide a clean and healthy environment.

2.1.2 Disease Transmission

Diseases causing germs come out of the infected human body (through discharges or contact) and mix with drinking water, food, milk, soil and air. Diseases are transmitted from person to person either through a vehicle (media like water, air) or a vector (like fly, mosquito). Depending upon the mode of transmission and the types of diseases, the following classification can be made.

1. insect borne (malaria, filaria)
2. water borne (cholera, typhoid, dysentry, diarrohea, jaundice)
3. air borne (influenza, tuberculosis)
4. food borne (enteritis, botulism)
5. virus borne (flu, poliomyletis)
6. respiratory (tuberculosis, bronchitis)
7. intestinal (cholera, typhoid)

Housefly mechanically transfers germs from filth to food that is left uncovered. The infected food, when consumed, cause severe dysentery, cholera and gastroenteritis. Anopheles mosquito (causative organism-plasmodium vivax) takes the blood of an infected person, undergoes a biological cycle, in her body, (converts gaemotocytes into sporozoites), and induces malaria parasites into a healthy individual's blood on her next feed. Filaria, caused by culex variety is an endemic disease. Amoebiasis spreads due to improper sanitation facilities. Brucellosis (from raw milk), Salmonellosis (from poultry and meat products) and trichinosis (from pork) are not uncommon when the animals are infected. Cholera and plague are dreadful diseases. Environmental component plays a major role in passing on the disease causing organisms from an infected person to a healthy individual. Timely immunization is the only way to fight some of the dreadful diseases which often prove to be fatal. It is possible to immunise through vaccination for polio, tetanus, diphtheria, cholera, hepatitis, tuberculosis, etc. as a protection against those diseases. Sick persons may be isolated (quarantine measures) and their discharges, disinfected. But in a poor, illiterate, developing country it may not be quite possible to take the above measures with maximum success. The best solution lies only in keeping the environment clean by maintaining proper sanitation measures. Health education creates awareness among the general public. It is the duty of a public health engineer to protect the surroundings and prevent the communicable diseases. Diseases like fluorosis which is related to drinking water, can be controlled by supplying of water with the prescribed level of fluorides.

2.1.3 Principles of Sanitation

Sanitation is not a simple task. It involves the coordinated efforts of various agencies to maintain a clean surrounding and prevent spreading of diseases. All the factors that are responsible for transmitting diseases, must be controlled at a time without fail. It needs an integrated approach to promote health and happiness in a civilised society.

Some of the essential steps for controlling the transmission of diseases are as follows:

1. Provide suitable living conditions in the human habitat (including good housing facilities).
2. Effectively collect and dispose solid wastes from residential units and surroundings.
3. Control pests and insects such as flies and mosquitoes, and eradicate rodents from the domestic environment.
4. Protect food and milk products from contamination during the various stages of preparation, storing, display or preservation. Food products should be kept covered and free from insects, microorganisms and toxic chemicals or adulteration.
5. Supply safe and wholesome water for drinking and domestic use.
6. Collect, treat and safely dispose sewage from residential areas.
7. Educate farmers regarding hookworm and other diseases which are caused due to exposure to contaminated soil.
8. Prevent stagnation of water in low lying areas. Use disinfectants, if required.
9. Create safe working environment inside the factories for preventing occupational hazards.
10. Safeguard public health during camps, festivals, or at places where people meet in large numbers.
11. Maintain sanitation in hospitals, slaughter houses or at places where infectious organisms grow and multiply.

Slum clearance, underground drainage system and protected water supply schemes are essentially needed for the welfare of a community. Observations have revealed that the mortality rates can be drastically reduced if purified drinking water is supplied.

Safeguarding public health has become more of a combined team effort of a civil engineer, in collaboration with town planners, microbiologists, chemists, medical practitioners and social workers.

2.1.4 Vector Control

Uncovered food items, stagnated water, overflowing garbage bin, leaky taps and drainage pipes, open manholes, unprotected drain outlets, exposed faecal matter, are the places for insects to breed and multiply. The following precautions should be taken to control the menace caused by various insects and pests.

Fly control

House fly (musca domestica) carries infectious microorganisms from excreta and contaminated food articles and mechanically transmits diseases such as cholera, typhoid, dysentery, and gastroenteritis. Fly control measures are given below.

(a) Keep the surroundings always clean and dry.

(b) Eliminate the breeding of flies by disposing human excreta through water carriage system.

(c) Dispose animal dung or decomposed food articles away from living areas.

(d) Prevent the flies by storing all food articles in proper containers. Cover the food properly and keep in a wooden safe fitted with wire mesh.

(e) Prevent easy entry of insects into the house by screening doors and windows with a fine mesh, allowing the doors to open outwards and adopting double door system as shown in Fig. 2.1.

(f) Use insecticides to kill flies in larvae and adult stages. Borax, gammaxene, malathion, pyrethrum, chlordane, and sodium arsenate are some commercially available products.

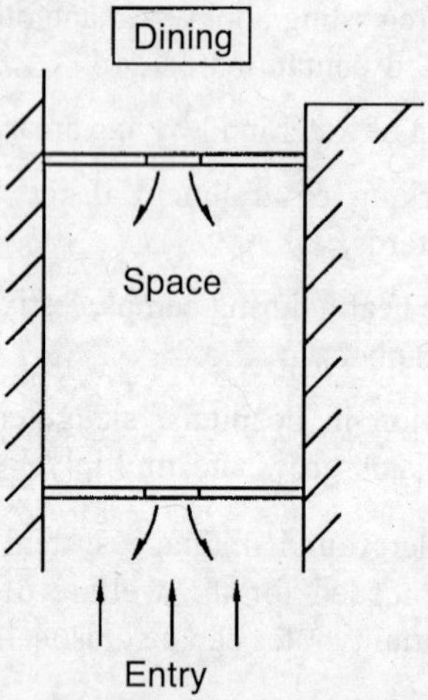

Fig. 2.1 Double doors.

Mosquito eradication

One should know the various stages during the breeding process before embarking on an eradication drive. The following steps will prove to be effective in eliminating mosquitoes.

(a) Mosquitoes breed on stagnant waterpools and shrubs in the domestic surroundings. So, remove shrubs and vegetation, construct subsoil drainage systems, fill up low lying areas and pits and cover the street drainage to eliminate breeding places.

(b) In water bodies (reservoirs), mosquitoes lay their eggs and larvae cling to the top surface water layers. So flooding or fluctuating water levels in reservoirs kills mosquitoes at the larvae stage. Moreover, Gambusia type of fish in water bodies eats larvae.

(c) Use insecticides to kill mosquitoes in larvae and adult stages. DDT, dialdrin, pyrethrum, parisgreen, malariol, lindane are some products available in the market.

(d) Fitting doors and windows with a fine mesh prevents the entrance of mosquitoes into the dwelling units.

(e) Use of bed nets and repellants also help an individual against mosquito bites.

Rodent control

Ratfleas are carriers of plague causing microorganisms. Rats destroy valuable property and eat food articles on a large scale. In the fields also, rodents destroy sugar cane and other crops. Rats live in the inner side of thatched roofs, in secluded corners, and in burrows in courtyards or fields. They live on anything, under any circumstances and are clever enough to escape easily.

Some methods of exterminating rats are given here.

(a) Prevent rats from reaching out to food items, water and cupboards/boxes by keeping these things covered and closed when not in use.

(b) Eliminate entry of rats into residential areas by using cement, stone, brick and other pucca materials during construction of houses.

(c) Plug all the house drainage outlets with heavy mesh-gauge coverings.

(d) Rats walk along the wall sides on selected routes. So an obstruction in the shape of a curtain wall (Fig. 2.2) constructed around the foundations, prevents entry of rats into the house to a great extent.

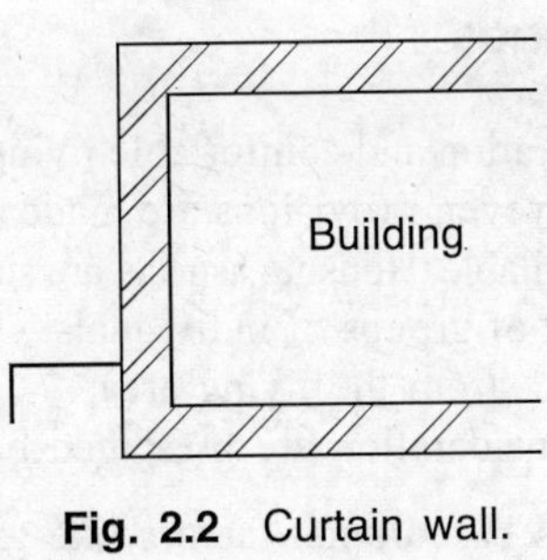

Fig. 2.2 Curtain wall.

(e) Trap the rats carefully with suitable baits, as they are very elusive.

(f) Kill the rodents with poisons like arsenic, barium carbonate, alpha naphthyl thiourea (ANTU), sodium fluoroacetate, thallium sulphate or zinc phosphide.

(g) Fumigate rat burrows in the fields (outside residential units), using gaseous form of chemicals such as sulphur dioxide or hydrocyanide gas.

Use of chemicals

A number of chemicals, in solid, liquid or gaseous states, are used to kill rats, insects and microorganisms. When they are used in houses, care should be taken to protect food articles, pet animals and little children from exposure to the toxic effect of the chemicals. Selection of a disinfectant (larvicide, insecticide, fungicide or rodenticide), use of proper dosage and method of application require expertise and continuous monitoring. The following points are of general interest:

- Sunlight and fresh air are natural disinfectants.
- Microorganisms are killed by applying heat in various forms.
- Phenol, carbolic acid, cresol (belonging to coal-tar group) are used to disinfect sputum, excreta, floors, clothing.
- Bichloride of mercury is used for washing wounds, and containers.
- Copper sulphate destroys algae and microscopic organisms in water storage tanks.
- Lime has germicidal action.
- Discharges from a sick person are disinfected with bleaching powder.
- Chlorine kills harmful bacteria present in water tanks.
- Sulphur, pyrethrum, arsenic and sodium fluoride are used in solid state to kill insects.
- Carbon disulphide and carbon tetrachloride are applied in liquid form.
- Sulphur dioxide, phenol, methyl bromide, carbon disulphide and HCN are used as fumigants.

2.1.5 Housing Needs

Home is a place for relaxation and comfortable living. Basic housing requirements are common for all. However, provisions are made depending upon the standards of living and finances available. Housing aspects are supposed to satisfy physiological and psychological needs of groups of individuals. Good housekeeping eliminates insects, vermin, rats, etc. from the living area.

Some points for consideration are presented below:

1. Site selection is based upon factors like ground elevation, accessibility and proximity to market, school, bus stand, railway station and hospital. But it should be away from busy, crowded and noisy environment.
2. Building orientation is planned in such a way as to get the best advantage of natural features like fresh air and sunlight for warmth and visual comfort. Depending upon the need, artificial illumination and ventilation

are supplemented. Open space (courtyard) of 30% of the total land area, window area of 20% of total floor area, and more than 3 m roof height are sufficient to provide proper ventilation.

3. Other requirements are
   (a) Number and size of rooms to provide comfort for the family
   (b) Smoke free kitchen
   (c) Water, toilet and drainage facilities with good plumbing fixtures
   (d) Strong walls and pucca construction (weather resistant and fire proof)
   (e) Basement height of 1 metre above the ground level

People generally reside in small units in the heart of cities or towns, whereas rural population have more living space per person. In low income groups, living area per capita is drastically less and common facilities are shared.

2.1.6 Community Sanitation Measures

Whenever and wherever people meet in large numbers either in a camp or at a festival, there is every possibility of transfer of diseases from person to person due to overcrowding.

Communicable diseases are prevented by adopting the principles of sanitation. Competent and regular supervision is very essential regarding various aspects of water supply, drainage, provision of latrines, refuse disposal, food poisoning, immunisation schemes, etc.

Sanitation measures are outlined as:

1. Control all points of entry and exit at the camp or festival site.
2. Avoid overcrowding if there is not enough open space.
3. Immunise all persons visiting a camp or workshop against diseases like cholera.
4. Provide adequate facilities for water supply, toilet, bathing, housing, etc.
5. Allow proper surface drainage to avoid stagnation of water or waste water.
6. Collect and dispose solid refuse matter safely at regular intervals.
7. Chlorinate all wells and other drinking water sources.
8. Ensure proper disposal of faecal matter.
9. Use lime, bleaching powder, pyrethrum, etc. to control vectors.
10. Prepare, store, display and serve all food and eatables under hygienic conditions. Use of single service containers is preferred.
11. Create temporary accommodation shelters with good ventilation, lighting, living space and clean surroundings.
12. Take all steps to prevent accidents, fire hazards, drownings, etc.
13. Provide first aid, emergency and medical services and fire fighting equipment in readiness.
14. Employ skilled, trained and dedicated personnel for various jobs.

In fact, sanitation is not a stop-gap arrangement. Unless total attention is continuously bestowed on all the aspects, results cannot be achieved. On a similar note, a list of measures to be taken in schools, hospitals, hostels, slaughter houses or swimming pools can be outlined.

2.1.7 Health Education

Health education creates awareness and plays a vital role in all community health programmes. Publicity through pamphlets, lectures, slogans or posters using newspapers, periodical magazines, radio and television network, is the most effective means to enlighten the masses about basic health rules. Incentives can also be given to communities that implement better sanitation measures and maintain cleanliness.

Publicity materials can be prepared on the following aspects:

1. The damage caused by disease
2. Source of disease and the method of disease transmission
3. Preventive and curative methods
4. Actions to be taken by individuals and communities to prevent transmission of diseases.

The amounts that are spent on this type of health education pays rich dividends for the society on the long run.

2.2 OCCUPATIONAL SAFETY

2.2.1 Hazards in Various Types of Occupation

A person works for about 8–10 hours a day in a factory unit, involving himself with machines, process chemicals and generated wastes. His operations, exposure and associated problems depend upon a variety of conditions. It is not an individual problem. Hence occupational welfare comes under community health plans. Some problems and solutions are indicated below.

1. Mechanical, monotonous and repetitive working operations cause physical strain. This can be eased by convenient and suitable arrangement of devices and proper work distribution schedules.
2. Heat, cold, humidity, light, noise, vibrations, pressure, compressed air or radiation affect the workers in the working atmosphere. Workers may be provided with personal protective devices.
3. Biological infections like brucellosis, tetanus, anthrax originate from tannery, dairy, poultry and slaughter house wastes. In addition to sanitation measures, preventive and curative medical care is necessary to arrest infectious diseases.

4. During the transportation, storage, and handling of toxic chemicals, combustion, explosions or corrosive reaction may take place. Safety precautions and procedures help the individual handling the toxic material.
5. Lead, mercury, zinc, arsenic, chromium, silica, beryllium, asbestos, cadmium, carbon monoxide, chlorine, sulphur dioxide, hydrogen sulphide, hydrogen fluoride, ammonia, cyanide, carbon disulphide, benzene, vinyl chloride are some of the chemicals present in the environment depending upon the processing units. These chemicals react with human system (respiratory, digestive or nervous) and damage liver, lungs, kidneys, blood, brain or other parts. Allergy, asthma, breathlessness, conjunctivitis, skin disorders are also very common to workers in various industries.
6. Leakage, breakdown and mechanical failure may lead to accidents causing burns and injuries to workers. Sometimes fire, explosion and inundation may also occur in industries and mines. On all such occasions, emergency management plans that include preventive measures and regular maintenance can only save the situation.

2.2.2 Objectives of Occupational Health Plan

The main objective is to improve the conditions at the work site and maintain high degree of physical, mental and social health of workers in the community. This is always advantageous to the industry also, as the productivity can be maximised only in an environmentally healthy atmosphere.

The following steps are useful:

1. Periodically monitor the environment within the factory. A regular health check-up of the worker will establish to what extent the worker's environment is responsible for the ill health (if any) of the worker. Monitoring helps in identifying the potential dangers in the work unit. It also helps in keeping the toxic substances at the work site within the permissible levels of exposure. If the health of any worker is still found to be affected due to exposure to toxic substances, the person is referred to the curative medicine department.
2. Strictly adopt all precautionary procedures. This may be with regard to light and ventilation aspects near the place of work or in connection with wearing helmets, goggles, respirators, etc. In any case, work culture and code of conduct must be maintained with utmost care.
3. Prepare a check list of various machines/process operations for regular inspection and maintenance. Preparing standard proforma for each unit in a factory helps in an easy and thorough verification of the maintenance without any omission. Upkeep and maintenance of the machines is not the responsibility of one person; it should be a part of the daily routine in any organisation.
4. Create awareness and impart training to the workers in the concerned

fields of specialization, at regular intervals. The working skills also need to be improved, from time to time. Occupational hazards and preventive measures must be known to the workers.

5. The ultimate objectives of any occupational health plan are to control and eliminate all occupational hazards. First, one has to identify the potential sources (causes) of risk and danger, and then adopt all preventive care in the true spirit to eliminate them.

2.2.3 Prevention and Control

It is not one time, one man and one unit problem. It is not just connected with diseases that are likely to attack a group of working personnel. It is not even connected only with the possibilities of accidents and safety provisions, provision of exhausts and strong foundations for machines. The chief aim is to improve all working conditions to increase the overall efficiency of a worker. The height of a bench, the position of a leg, the body posture, the periodicity of lifting a hand, etc., all play a major role in the worker's operational schedule. So, occupational health measures are for the total improvement of the working conditions, worker's welfare and efficiency. The following health measures covering various aspects of work environment are given below.

General precautions

(a) Try to use the dangerous chemicals to the minimum possible extent.

(b) Avoid direct contact between the worker and the source of hazard.

(c) Good housekeeping, floor maintenance, repair and upkeep of machines and moving parts, prevention of leakages and spillage are included in the plant sanitation. Adequate provision of drinking water systems, toilet, canteen, rest room, education and recreation facilities should also be provided to the workers.

(d) Education and training to the workers regarding the use of personal protective devices like aprons, goggles, gloves, ear plugs, helmets, etc. should be compulsory as it is quite useful. Knowledge of first aid should also be imparted as a part of training programme.

(e) Management should implement all the legal provisions as per various Acts of the government, such as Factories Act, Workmen Compensation Act, Hazardous Chemicals Handling Act, Employees Insurance Act, Environmental Protection Act.

Engineering aspects

(a) Site selection, building planning, foundations and structural safety are important considerations.

(b) Location of equipment and storage of chemicals need special mention.

(c) Ventilation and illumination are to be provided to the required extent in the working area. Exhausts are essential to remove the foul gases.

(d) Proper enclosures to segregate the work area from other zones.

(e) Dust elimination and noise control.

(f) Electrical and welding operations.

(g) Ergonomics (study related to the operations and movements of man with machine in a working environment).

(h) Environmental monitoring and pollution control are all part of the engineering aspects to improve the occupational health.

Medical aspects

(a) Periodical examination of the workers would reveal the presence of diseases related to the occupation.

(b) For preventive and curative therapy a medical practitioner, specialised in the field of industrial hygiene, should be consulted.

(c) Counselling and rehabilitation, whenever needed, instill courage and confidence in the worker.

(d) There should be facilities for shifting the personnel from a specified work area, if he reaches a certain permissible limit of exposure of hazardous substance.

(e) Industry should bear the entire costs in the treatment of occupational hazards.

Safety services

Safety comes from man's mastery of his environment and himself. It is won by individual effort and group cooperation. It can be achieved only by well-informed, alert, skillful people having regard for the welfare of others.

Simple safety precautions are:

(a) Warning signals and posters must clearly indicate the do's and don't's in that specified job area.

(b) Adequate personal protective devices should be used without fail.

(c) First-aid knowledge is a must for all workers, and the necessary kits must be available within the reach for immediate use.

(d) Emergency operations are required to meet the fire and explosion hazards, flood situation, electrical short-circuiting, accidents, etc.

(e) It is essential to close or shut down a particular unit when there is an imminent danger for an accident. Besides, workers must be trained to take remedial measures as instructed in the visual display boards kept at a number of places in the work place.

(f) All safety measures should be inspected at regular intervals of time. They should be in working condition in any emergency situation.

In conclusion, occupational health and safety depend upon the concentration levels of toxic chemicals in the environment, the duration of exposure, the sustainability of the individual and the remedial measures available in the industrial unit of operation. Workers' health is industry's wealth.

2.3 REFUSE DISPOSAL

2.3.1 Constituents of Refuse

The word *refuse* means solid waste. There are two components of refuse— *garbage* or food waste and *rubbish* or the combustible and noncombustible portion of solid waste.

- Garbage (kitchen waste), consists of putrescible or decomposing organic matter such as vegetables, fruits, food material, etc. Per capita kitchen waste is assumed to be 0.10 kg/day.
- Rubbish, consists of non-putrescible dry matter including glass, metal, leather, textile matter, paper packing material, house hold waste, etc. Per capita contribution is estimated as 0.25 kg/day.

It is always difficult to separate various constituents from a normal mixed refuse of a dust bin. Sometimes dead animals and animal dung also get mixed up with solid wastes. Building materials used in construction industry may also be dumped along with the normal refuse. Ashes consists of matter after burning.

Solid wastes may be classified as market refuse, stable refuse, street refuse, etc. They may also be identified as municipal, industrial and hazardous wastes, and their constituents may vary from place to place.

2.3.2 Problems after Disposal

Indiscriminate disposal of solid wastes is the source of putrefying offensive odours, unsightly surroundings, breeding of insects, disease transmission, and fire hazards. Solid wastes are scattered due to wind, water, animals and rag pickers. Refuse from industries may create more problems due to the presence of hazardous substances.

2.3.3 Refuse Collection

Bulk content can be reduced by cutting, breaking, folding, flattening or shredding operations and the process of compaction during collection.

1. RCC, dustbins (2-m height and 1-m diameter) located at street corners store solid wastes before collection and transportation by vehicles to the place of disposal.

2. Transport lorries should have mechanical arrangements for loading, compaction, unloading and cleaning operations. Otherwise, manual operations can be carried out. Compaction if necessary, may be done at the collection point and at the disposal site. Vehicles should be covered during transportation.

3. Location, size of the dustbin and frequency of collection are interrelated. Proper planning is therefore required to avoid insanitary conditions.

4. Route schedule for transportation is planned in such a way that each truck is loaded to its maximum capacity on each trip while reaching the ultimate dumping site and reduce the cost for carrying 1 tonne/km. This is one method of saving on fuel costs avoiding idle runs of a truck. The following points have to be studied before fixing a collection route (A) or (B and C) in Fig. 2.3.

   (a) location of refuse bins
   (b) estimation of volume at each collection point
   (c) time of travel
   (d) distance of travel
   (e) ultimate disposal site conditions

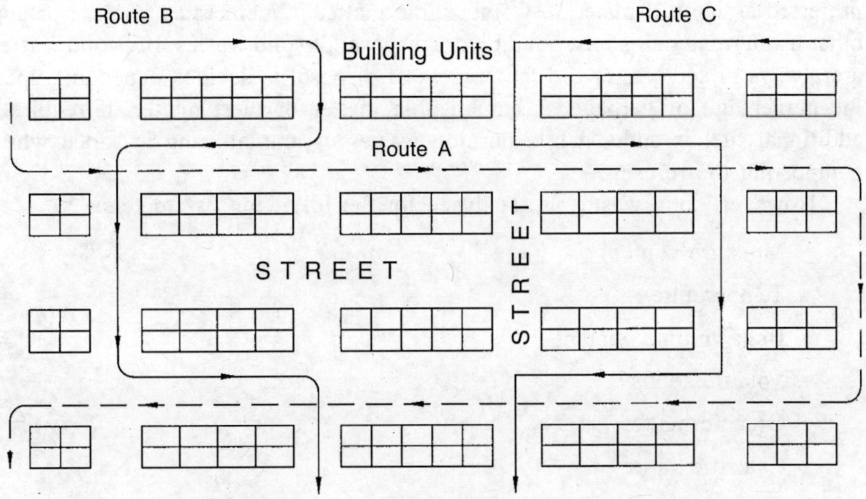

Fig. 2.3 Refuse collection route.

5. Separation of various constituents of refuse that consists of putrescible, combustible and inert matter is not easy and not advisable. But segregation is practised to recover some useful material from refuse. It is desirable at the first stage itself, for example, not to mix up a hospital waste with a market waste.

Plastic bags may be separated as they are not biodegradable. Magnetic material, and materials like paper and cardboard, are separated for resale in the market. In India, rag pickers pick paper, plastic covers, glass bottles, etc. from municipal refuse.

2.3.4 Recovery

Refuse is also treated as a resource if valuable material such as paper, plastics, metals, bottles, etc. are recovered before disposal of the mixed refuse. The refuse can be converted into manure or treated properly to produce heat.

1. Low lying lands are reclaimed by dumping the solid inoffensive wastes on land.
2. Methane (fuel gas) and manure are obtained from composting (digesting) organic wastes.
3. Heat generated during incineration process can be utilised.
4. Grease is having a resale value.

2.3.5 Parameters of Analysis

Refuse quantity and characteristics vary due to community habits, standard of living, place of collection, frequency of collection, season and climate. The purpose of analysis is to decide on the method of disposal. For example, organic wastes are used in composting process. Inorganic, noncombustible matter is preferred as landfill material. Combustible matter is incinerated. Grease or any other useful material is recovered before disposal of solid wastes. But solid wastes are always mixed wastes. So the process of disposal is decided depending upon the percentage of garbage or combustible matter or inert matter. If required, additional fuel is added while burning wastes or animal dung is added while composting the refuse.

However, solid wastes are analysed for the following parameters:

- Moisture content
- Temperature
- Bulk volume content
- Volatile solids
- Organic matter
- Calorific value
- Carbon to nitrogen ratio
- Grease, potash
- Ash content

Moisture content and temperature indicate qualities of putrescibility and bacterial activity. Bulk volume is useful to know the compressibility. Organic matter, carbon to nitrogen ratio, and moisture content are some useful parameters in the composting process. Calorific value is related to the fuel requirements. If grease, potash, etc. are found, recovery is suggested. Refuse can easily be disposed off in landfill operations if ash content inert matter) is more.

2.3.6 Landfill

Solid wastes are placed in sanitary landfill system (trenches, pits) in alternate layers of 80 cm thick refuse, covered with selected earth fill of 20 cm thickness, as shown in Fig. 2.4. After two to three years, solid waste volume shrinks by 25–30% and the land is used for parks, roads and small buildings.

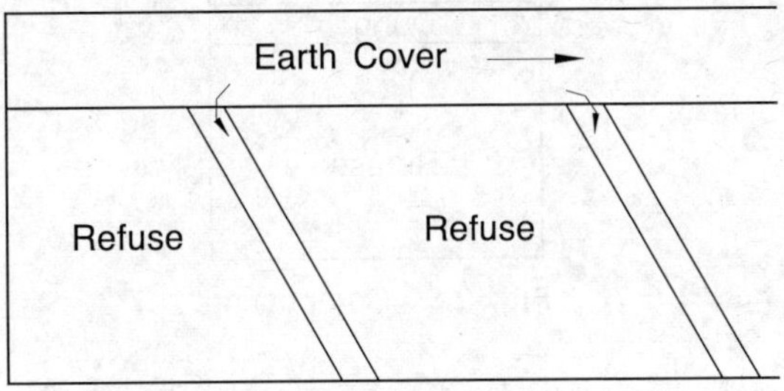

Fig. 2.4 Landfill.

With normal compaction, a landfill site can take 500 kg of refuse per cubic metre of trench space available. The method is simple and economical. Mixed refuse can be disposed without any segregation of constituents. Animal dung, human excreta or dead animals should not be disposed off in the landfill method.

Care should be taken while locating the site for dumping refuse. It depends upon availability of land area, soil conditions, ground water table, topography, distance from the residential area and ultimate usage of site after reclamation.

Disadvantages

(a) Bad odours, if landfills are not properly dealt with

(b) Uneven settlement of land

(c) Leaching of waste water and chemicals into the adjoining areas leading to soil and water pollution, which can be prevented by
  (i) proper surface water disposal and
  (ii) collecting leachate at the bottom of the landfill.

2.3.7 Composting

In the composting process, bulk organic waste is converted into a fertilising manure by biological action. Alternate layers of refuse 30 cm thick and animal dung 5–6 cm thick are filled in 1 m deep trenches of size 10 m × 3 m, as shown in Fig. 2.5. Finally, a 5 cm thick earth layer fills the top level, which prevents access to insects, loss of moisture content and public nuisance. In a period of 3 to 4 months, manure is ready for use.

In aerobic window composting method, it is exposed to the atmosphere. Frequent turning and control of moisture content helps in rapid decomposition of organic matter. In anaerobic process, intensive biological action takes place and

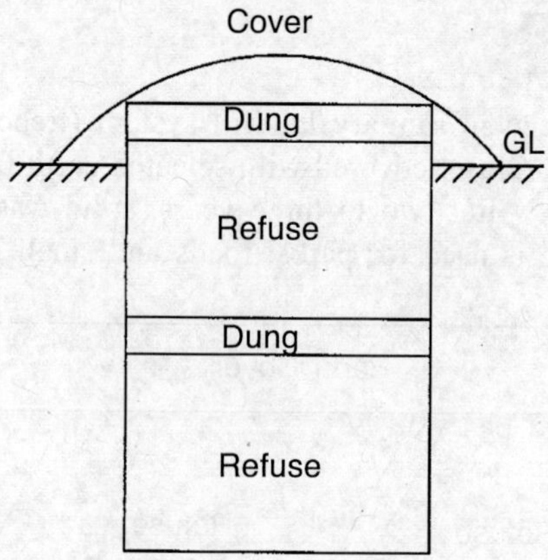

Fig. 2.5 Composting.

higher temperatures are induced inside the trench. Composting is also practised with the help of earth worms (vermin composting).

The following conditions are desirable during the composting process

1. Initial shredding of refuse
2. Removal of non-degradable matter
3. Seeding with partially decomposed matter
4. Maintaining 30 to 40% moisture content which is optimum
5. Carbon to Nitrogen content of around 30 (to 40):1

All these improve the rate of decomposition and quality of manure.

2.3.8 Incineration

Burning of combustible refuse is the safest process of disposal. Incineration is quite effective when the garbage content is 50–60% of the total refuse and calorific value is more than 1600 kcal/kg. Incinerators can be located at different parts of the town, to reduce the cost of transportation of refuse. Sometimes incinerator needs additional fuel or proper air supply to maintain high temperature in the burning place. Figure 2.6 outlines an incinerator. Hospital wastes may be safely disposed by the process of incineration.

2.3.9 Plastics

Rigid or flexible type of plastic containers are liberally used for packaging in consumer market. The material is polyvinyl chloride or polyethylene tetra phthalate (PET). They have long life and are not biologically degradable. The best way is to recycle and use to the extent possible. Fine threads or ropes can be prepared from used polythene covers. Plastic containers can be made into pads or panels

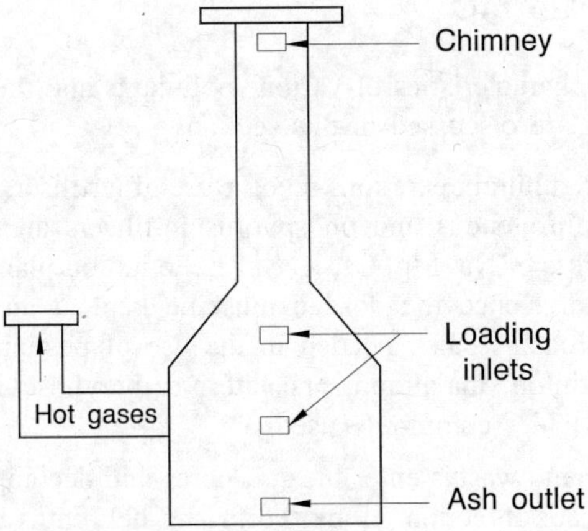

Fig. 2.6 Incinerator.

after proper melting and moulding. Use of plastic bags or covers is banned at some places. Research work should be carried out on a priority basis to recycle and reuse plastic material and find safer methods of disposal.

2.4 SOIL AND AGRICULTURAL POLLUTION

2.4.1 Top Soil

Land is a resource rich with minerals, nutrients, oil products, and natural gas. While the top soil is fertile and useful for growing crops and plants, it is greatly affected by soil erosion and sedimentation caused by wind and water. Storm, flood and runoff take away the top layer of soil and transport the same from place to place, thus making the rich soil lose its nutrients and minerals. Construction and mining activities result in loss of land for agricultural production. In urban areas refuse disposal activities allow the leachates to join the ground. In rural areas more of fertilisers and pesticides are used in agricultural operations and the residual chemicals remain in the top layers of soil. Addition of fertilisers, nutrients and organic matter may increase the soil fertility. Toxic insecticides kill useful soil bacteria favourable for plant growth. Land acts like a filter in removing the impurities in water and waste waters. But toxic residual chemicals from the soil reach human beings through fruits, vegetables and other food items. In fact, the concentration of chemicals increases en-route (by biomagnification) and is more dangerous.

Industries discharge liquid and solid wastes with or without adequate treatment which get deposited on the land or in water bodies. Thus soil and water resources become polluted.

Formation of top (rich) soil is a slow process. Replenishment takes considerable time. Soil is self purifying provided the pollution is within the limits and sufficient time is given for recovery. Otherwise soil contamination leads to land sickness and ultimate loss of land for ever.

2.4.2 Pollutants

The special characteristics of various pollutants and their harmful effects on the environment are discussed in this section.

1. Agricultural operations involve use of fertilisers and pesticides. A number of nitrogenous and phosphorus fertilisers are available in the market. Excess of Na, Mg, Ca, K, S, Zn, Fe inhibit plant growth and reduce crop yield. Hence the dosage must be kept at an optimal level. Cautious approach is also needed in the use of pesticides. Endrin, endosulpha, diazinion, malathion, arsenates, carbaryl, sevin, etc. are some of the pesticides commonly used.
2. Organic wastes enter the soil pores and decompose. Pathogenic bacteria spreads infection. Hook worm and helminths also cause diseases.
3. Compounds containing arsenic, mercury, chromium, nickel, lead, cadmium, zinc and iron are toxic to life. Fluorides also affect the plant development.
4. Water logging and salinity increase the dissolved salt content in ground water and also the soil. Some plants are very sensitive to soil pH and salinity. So land becomes unfit for irrigation due to water logging conditions.

2.4.3 Parameters of Soil Analysis

Soils may be analysed for the following parameters to understand the nature of soil and extent of pollution.

1. Soil classification, porosity, permeability and particle size
2. Moisture content, dissolved salts, acidity, alkalinity, pH, hardness
3. Organic and inorganic substances
4. Biological activity
5. Soil nutrients like N, P, K.

Fruits, vegetables and crops should also be tested for the residual chemicals as their quality depends on soil pollution.

2.4.4 Remedial Measures

The following remedial measures are suggested:

1. Understand soil, water, plant relationship and protect the mineral cycles of fixation. Nitrification is an indication of soil fertility.
2. Treat sewage before land disposal.
3. Preserve and protect top fertile soil. Control soil erosion by proper tree plantation measures.

4. Prevent entry of leachates into soil layers.
5. Carefully use fertilisers and pesticides or use alternate natural methods. Use of excess fertilisers results in loss of micro-nutrients, and frequent use of pesticides eliminates soil friendly bacteria.
6. Plough or mix the soil to improve aeration, porosity and permeability.
7. Rotate the crop pattern to allow the soil replenish the nutrients or leave the land uncultivated for some time.
8. Cultivate grass and grow papaya trees only, on polluted waste lands.

2.5 NOISE CONTROL

2.5.1 Normal Sound Levels

Noise is unwanted, especially because unpleasant and loud sound, causes nuisance and disturbance to the receiver. Normal sound levels are indicated (in decibel units) as follows:

Soft whispers	:	40 dB
Residential areas	:	45 dB
Normal conversations inside offices	:	50–60 dB
Libraries and hospitals	:	50 dB
Hotels, schools	:	60 dB
Commercial areas	:	70 dB
City traffic	:	70–80 dB
Inside foundry	:	90–100 dB
Siren	:	100 dB
Blowers and drilling	:	more than 100 dB
Aircraft	:	100–110 dB
Hammerfall	:	120 dB

2.5.2 Ill Effects

Noise interferes with normal human activities. It has an adverse effect on the hearing ability. Continuous exposure to loud noise leads to physiological disorders or psychological stress. Irritation, frustration, fatigue, headache, anxiety, hypertension, insomnia, hearing loss, mental depression are generally associated with exposure to high intensity of noise for short durations or moderate levels for a prolonged period. Discomfort and lack of concentration caused by noise affect the worker's efficiency and even lead to accidents. Sudden explosions create cracks in building walls and break window glasses.

It is observed that the effect of high intensity sound can be damaging for our physical as well as mental health.

(a) 80–85 dB sound causes discomfort,

(b) 100–110 dB sound causes stress,

(c) 120–130 dB sound causes nausea, dizziness, pain in the ear due to prolonged exposure and

(d) >150 dB sound causes permanent damage to the hearing mechanism.

These effects of noise pollution depend upon (i) the distance of the observer from the noise of generation, (ii) the duration of exposure to that intensity of sound level, and (iii) combined effect due to multiple sources producing different sounds.

Acceptable noise levels are:

- in silence zone (schools/hospitals) : 45 dB
- in residential areas : 55 dB
- in commercial areas : 65 dB
- in industrial zone (working environment) : 75 dB

2.5.3 Noise Measurement

Sound waves propagate as longitudinal waves in an elastic medium. Sound intensity is the rate at which energy passes through a unit area normal to the direction of movement. It is related to sound pressure level (L) as 10 log $\frac{I}{I_o}$ where I = measured sound intensity (watt/m^2) and I_o = reference quantity (20 micro pascals or 10^{-12} watt/m^2). This reference quantity (the softest sound level that could be detected) is a standard threshold of audibility at frequency of 1000 Hz, equivalent to zero decibel.

Frequency of sound is the rate at which air pressure varies, i.e. number of cycles per second (hertz). Loudness (an auditory perception of the observer) depends upon the effective frequency and acoustic pressure. It is expressed in sones as 1 sone = 40 dB at 1000 Hz. Same intensity but different frequencies produce different levels of loudness depending upon the ear sensitivity. As sound levels are measured on a logarithmetic scale, 50 dB sound for 30 minutes duration and 70 dB sound level for 20 minutes duration afterwards add upto an equivalent noise for 50 minutes duration level of 64 dB, i.e.

$$10 \log \left(10^{50/10} \times \frac{30}{50} + 10^{70/10} \times \frac{20}{50}\right) = 64 \text{ dB}$$

Sounds are recorded for further evaluation. Frequency analysers (with a constant bandwidth) have microphone with an amplifier and display. Sound level meters have a microphone and electronic circuits as weighing networks. These networks are classified as A, B and C. Very low frequencies are filtered out by the 'A' network. The sound level is lower than other networks (B and C). Readings are obtained in 'A' weighted scale and represented as dB (A). The advantage of these networks is that sensitivity can be varied to attenuate certain frequencies and intensity of sound is similar to that received by the human ear.

Graphic recorders are used to plot sound levels with reference to time. It is also possible to detect peak values, which cause hearing damage. Noise dosimeters are used to monitor the exposure of a worker in the factory.

2.5.4 Preventive and Control Measures

A number of provisions are made in Factories Act (for industrial noises), Motor Vehicles Act (for automobiles), and Public Nuisance Act (for using loudspeakers in a residential area). Various aspects of prevention and control of noise are discussed below.

1. Educate the people to follow legislation, according to which the level of sound from loudspeakers at parties, prayer halls and public meetings should not cross the prescribed limit.
2. Plan traffic movement for a smooth flow of vehicles.
3. Insist on the use of silencers on vehicles.
4. Create silence zones near schools, hospitals and prayer halls to prevent the use of loudspeakers, crackers and vehicle horns.
5. Rigidly fix the machines on concrete blocks and place absorbent material to reduce industrial noise due to impact and vibration. Steel springs, rubber, felt, and cork reduce transmission of mechanical vibrations. Absorption limiting materials such as asphalt also dampen the sound levels.
6. Construct acoustical cabin enclosures for various units of operations like drilling and grinding to arrest noise levels at source. Areas or rooms producing noise may be segregated so as to create a buffer zone. Sound-proof office rooms are desirable.
7. Ensure smooth working of machines with proper repair and maintenance such as rivetting the joints, using lubricants for machines etc.
8. Regulate time of exposure for the industrial workers and insist on the use of personal protective ear plugs.
9. Plant trees on road margins.

Compound wall, courtyard space, vegetation cover, also act as a sound barrier for the absorption and dissipation of sound energy.

SUMMARY

- Diseases are transmitted through vectors (flies, mosquitoes, rodents, etc.) and vehicles (water, air, food, etc.) in human environment. In a commercial, institutional and other public places, diseases may be transferred from person to person even in a healthy social environment. Environmental sanitation measures include supplying purified water, disposing of wastes, controlling insects, maintaining hygienic surrounding, etc.

- In industries and work places, people are exposed to occupational hazards caused by various physical, mechanical and biological agents, which can be controlled by various engineering, medical and safety operational procedures.
- Solid-wastes are a potential source of nuisance in a city. They should be safely disposed by a sanitary land fill operation or incineration.
- Pesticides and fertilisers used in various agricultural operations pollute top layers of soil and water sources through soil. Top fertile soil is protected by taking adequate precautions in cultivation practices, preventing water logging and salinity conditions, reducing soil erosion, and treating wastes before land disposal.
- Unwanted sound interferes with normal activities. Constructing sound barriers, enclosures and tree plantation, reduce the adverse impacts in the working environment due to noise levels.
- Sanitation (cleanliness) ensures safe living conditions and improves health and welfare of a community.

REVIEW QUESTIONS

1. Briefly explain the role of environmental components in disease transmission.
2. List out the various sanitation measures to be adopted in a community health programme.
3. Write a note on the measures taken for improving the health and safety of industrial workers.
4. Describe the merits and demerits of different methods of refuse disposal systems.
5. Explain in detail the measures adopted to prevent soil pollution.
6. Indicate suitable methods to prevent and reduce noise pollution.

OBJECTIVE TYPE QUESTIONS

Fill in the blanks with suitable words:

1. _______ is a water borne disease whereas tuberculosis is a _______ disease.
2. Health education creates awareness; One example is, advertisements regarding _______ (disease).

3. The breeding places of mosquitoes are________.
4. Fumigation is useful in ________.
5. Chlorination is a practice to ______.
6. Common chemicals present in the industrial environment are ________ and ________ and disorders common to workers are ________ and ________.
7. Traffic police man is likely to be exposed from ________ (chemical), which causes ________ (disease).
8. Composting is suitable for ________ type of refuse.
9. Disadvantages of water logging are ________.
10. Sound intensity levels are reduced by ________, as a measure to prevent noise pollution.

CHAPTER 3

Drinking Water

3.1 WATER QUALITY

Almost three-fourth (72%) of the earth's surface is covered with water. Although water is the most common liquid, it is also one of the most unusual. Water has certain special properties that have important consequences on the ecology of our planet. Water is the only substance that expands on freezing. This unusual characteristic (unusual variation in density in the temperature range from 4°C to 0°C) makes aquatic and marine lives possible during the severe winter months in the polar regions. The temperature difference between the freezing point and the boiling point of water is also very wide unlike other substances. It is also the only liquid that takes much longer time to get heated up and cooled. But this peculiarity of the physical and chemical properties of water is so much essential for sustaining life on earth. Little wonder then that early civilizations developed around water resources only, such as the river Sind and river Nile.

3.1.1 Sources of Water

Rain and glacier are the main sources of water. Water from these sources flows over the surface to form lakes, ponds, reservoirs, rivers, canals, creeks and are termed as *surface water*. The water which gets absorbed into the ground to be tapped as springs or wells, shafts or infiltration galleries, are termed as *ground water*. As water slowly percolates through the ground, the layers, through which it passes, act as a filtering material and have purifying effect on it. Therefore, deep well water is usually clear, colourless and low in bacterial content, although mineral content is generally higher. A good amount of rainwater percolates through ground layers. The top surface of saturated zone below which water is available is indicated by ground water table. Reservoirs generally store drinking water and supply regularly to different localities, while lakes are water bodies for supplying water to small towns. Rivers mostly give continuous water supply throughout the year. But some rivers have less flow in summer months. Besides lakes and rivers, water is available below ground level due to storage in subsoil water bearing formations. Ground water is collected by dug wells or tube wells depending upon the depth at which adequate quantity is available. Ground water yield depends on the rate of replenishment.

Although water is collected mostly from surface (through intakes) as well as underground sources (through pumping) water may also be collected from the rivers by the infiltration pipes and galleries. Only a very small percentage (less than 0.5%) of entire rain fall is available for human use since water from different sources ultimately flows down and join the ocean. Some water is inaccessible and some is lost through evapotranspiration. (See hydrological cycle in Chapter 1). The intensity, duration and frequency of rainfall at a place can to a certain extent be predicted from the previous years' records.

Quantity of water in a river is determined as product of flow velocity and the area of cross section of water in the river at any given time. When water flows through notches or weirs (Fig. 3.1), flow can be calculated using the relevant formula,

$$Q \propto KH^n$$

where Q = quantity of flow in m^3, H = head of water in the device in m, K and n are the coefficients depending upon the outlet conditions.

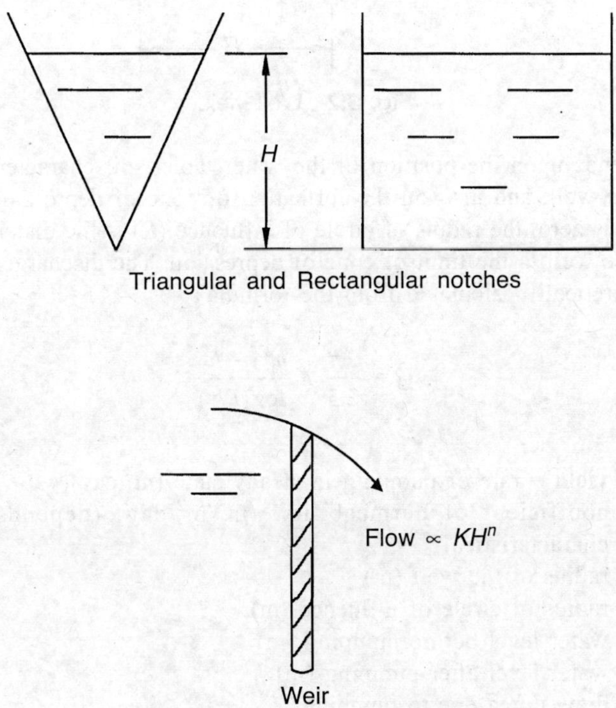

Fig. 3.1 Flow measurement.

3.1.2 Yield of a Well

Yield of a well is the quantity of water that can be obtained (in L/min or m^3/day) from a well under constant rate of pumping for a considerable time interval. Water level lowers (i.e. draw down) during pumping and an inverted cone of depression is formed as shown in Fig. 3.2. The quantity of water available from

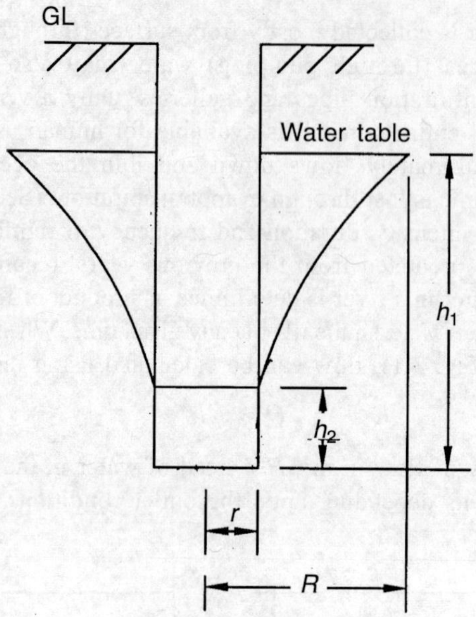

Fig. 3.2 Well yield.

a well depends upon the position of the water table, soil characteristics, water depth in the well, and also on the formation of cone of depression. When the draw down is zero, the radius of circle of influence (R) is the distance from the centre of the well to the limit of cone of depression. The discharge from a well can be theoretically calculated from the formula

$$Q = \frac{2\pi}{2.3} K \frac{h_1^2 - h_2^2}{\log(R/r)}$$

where

Q = yield = rate of pumping in steady state (m^3/day),
K = coefficient of permeability, (m^3/m^2/day) (depends upon soil characteristics)
r = radius of the well (m)
R = radius of circle of influence (m)
h_1 = water level before pumping (m)
h_2 = water level after pumping (m)
$h_1 - h_2$ = draw down due to pumping.

PROBLEM 1 Find the diameter (d) of a well, given that

Q (yield) = 12,000 m^3/day
K (coefficient of soil permeability) = 50 m/day
R (radius of circle of influence) = 300 m
H (aquifer thickness) = 30 m
$h_1 - h_2$ (draw down) = 6 m.

Solution

$$Q = \frac{2.72\,KH\,(h_1 - h_2)}{\log\,(2\,R/d)}$$

or,

$$\log\left(\frac{600}{d}\right) = \frac{2.72 \times 50 \times 30 \times 6}{12{,}000} = 2.04$$

or,

$$d = 5.5 \text{ m}$$

Well yield is determined by pumping test or recuperation test in the field.

(a) In a *pumping* test, when water level remains fairly constant (without causing any further increase in draw down) during pumping, pumping rate is taken as the yield.

(b) In a *recuperation* test, pumping is stopped after some time and the time taken for the water to regain the original level (or a certain height) is noted. The discharge from a well is calculated as

$$Q = 2.3\,\frac{A}{t}\,\log\left(\frac{h_1}{h_2}\right)$$

where

Q is the specific yield per unit draw down (m^3/h)
h_1 is the depression head (m) in the well after pumping is stopped.
h_2 is the depression head (m) after time interval t(hours)
A = ($\pi d^2/4$) area of the well (m^2), with diameter d(m)

PROBLEM 2 During a recuperation test, the draw down decreased from 4 m (h_1) to 3 m (h_2) in 2 hours (t). Calculate the specific yield of a well Q, when the diameter (d) of the well is 4 m.

Solution

$$Q = 2.3\,\frac{\left(\pi\,\frac{d^2}{4}\right)}{t}\cdot\log\left(\frac{h_1}{h_2}\right) \text{ m}^3\text{/h.}$$

$$= 2.3\,\frac{(4\,\pi)}{2}\cdot\log\left(\frac{4}{3}\right) = 2 \text{ m}^3\text{/h.}$$

3.1.3 Impurities

Rain water collects dusts and gases from the atmosphere. Similarly, surface water, during its movement on the surface of earth, collects more of suspended, dissolved and colloidal matters, which may be organic or inorganic in nature. River water contains more dissolved oxygen.

Underground water is in contact with soil formation and dissolves more minerals and salts. Suspended impurities are likely to be less in ground water, as subsurface porous soil acts as a filter. Water quality depends upon the catchment area conditions in the case of run-off, and mineral rock formations in the case of ground water. Water quality from a storage reservoir appears to be good as suspended solids settle down and upper layers are exposed to more aeration. As more nutrients are available at the bottom of the reservoirs, they may stimulate algal growth. However, rubbish from the town, fertilisers and pesticides from agricultural fields, domestic and industrial effluents may join any water course and completely change water characteristics.

Depending upon their nature, impurities present in water may be classified as

- Physical, chemical, bacteriological
- Suspended, colloidal, dissolved
- Organic, inorganic
- Mineral salts, toxic chemicals

Water has unique characteristics with regard to density, melting and boiling point and specific heat. It is a good and universal solvent and dissolves substances when in contact. That is, water from any source is likely to have impurities in some form or the other and there is no pure water, practically, anywhere.

All substances present in water do not necessarily have harmful effects. In fact, some are essential to impart good potability to water. To be consumable, water should be

- uncontaminated and hence free from germs causing water borne diseases
- free from toxic substances and
- free from excessive amounts of mineral and organic matters.

To be potable, water should be significantly free from colour, turbidity, bad taste and odour and should be stored at moderate temperature during summer and winter. Water becomes impure when it contains one or more substances in excess concentrations, or toxic chemicals which might have adverse effects on the user. Impurities, generally found in water, are classified according to their characteristics.

Physical

(i) *Colour*. It is due to colloidal particles, substances in solution, decaying vegetation, organic matter, inorganic salts, dyes and other compounds. Dissolved substances impart true colour, whereas suspended solids cause apparent colour. Pure water is always colourless.

(ii) *Tastes and odours*. Decomposed organic material, minerals and salts of Fe and Mn, vegetation, phenols, chlorine and algae impart typical tastes and odours to water. They are unpleasant and undesirable. The amount of water (dilution) that is needed to neutralise the disagreeable odour conditions is taken as a quantitative measure (threshold number) of odour intensity.

(iii) *Suspended solids*. Surface run-off carries suspended solids into water sources. The suspended solids may be inorganic in nature, such as clay, silt, sand, silica and calcium carbonate or they may be organic matters, e.g. oils, fats or grease. The quantity of clay and sand carried by muddy rivers are very large, which often interfere in chemical reactions. Some absorb or react with the dissolved oxygen. Oil and grease obstruct absorption of sunlight. They also create problems in treatment units. Dosage of chemical (viz. chlorine in the process of disinfection) will be less when once suspended solids are removed in the earlier units of sedimentation and filtration. Suspended solids can be easily removed physically by settling, flotation or filtration methods.

(iv) *Turbidity*. It is caused by suspended and fine insoluble particles and colloidal impurities like clay, silt, algae and plankton. Turbid water does not allow passage of light which is either absorbed or scattered, depending upon the size and surface characteristics of particles. Turbidity causes undesirable tastes and odours, apparent colour, and affects the process of photosynthesis for algal growth in lakes and ponds.

Turbidity measurement is different from the measurement of suspended solids. The relationship between the two may be highly variable. Nephelometer measures the intensity of light scattered by the particles. Jackson's turbidity meter measures depth of visibility of a candle light through a turbid sample (Fig. 3.3). A graduated rod is immersed in surface waters to obtain turbidity level.

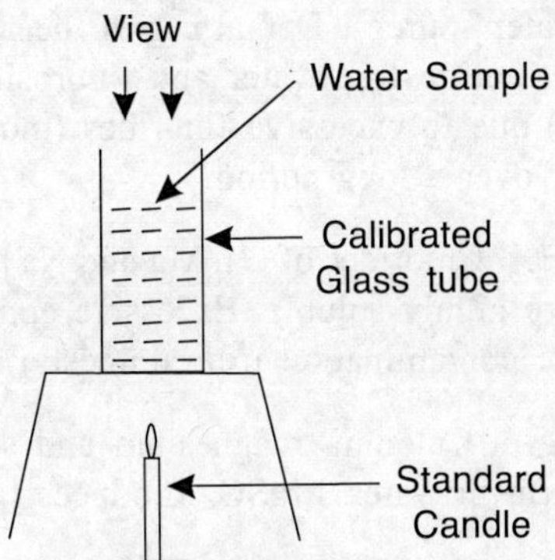

Fig. 3.3 Jackson's turbidity meter.

(v) *Temperature*. Palatability, solubility, viscosity, chemical reactions and odours are influenced by temperature. More quantities of chemicals get dissolved in water at higher temperatures, but gases (including dissolved oxygen) are expelled. Viscosity is reduced. Sedimentation and chlorination processes and biochemical oxygen demand (BOD) reactions are temperature dependant.

Chemical

(i) *Acidity, alkalinity and pH.* Acidity is caused by CO_2, mineral acids and salts from strong acids. Alkalinity is due to carbonates, bicarbonates and hydroxides. It is the capacity to neutralise acid. pH value denotes the nature and intensity of water towards acidity or alkalinity. pH is the Hydrogen ion concentration expressed as $\log_{10}[1/H]$. Since pH value of 7 indicates neutrality, 0–7 acidity, and 7–14 alkalinity, pH value of 6–8 is desirable. Bicarbonate ion is a component of both alkalinity and acidity. Aquatic life is more sensitive to pH. pH is important in chemical reactions and is significant in the treatment units like coagulation, softening, chlorination and sludge digestion.

(ii) *Hardness.* It is due to multivalent metallic cations (Ca, Mg) in solution, associated with anions, like sulphates, chlorides, bicarbonates. Hard water reacts with soap and prevent lather formation; and causes scale formation in hot water boilers in industries. So, hardness is to be removed for economical reasons. Hard water generally contains more than 150 mg/L of $CaCO_3$ whereas soft water may contain less than 75 mg/L. Hardness is classified as carbonate and non-carbonate hardness, Ca or Mg hardness, and temporary and permanent hardness.

(iii) *Total dissolved solids (TDS).* Water being a good solvent, dissolves a number of salts, chlorides, sulphates, bicarbonates of Na, Mg, Ca, K, etc. But all dissolved substances are not undesirable. About 500–1000 mg/L is good for domestic use, but excess concentrations are harmful in many ways. Some of these constituents are good nutrients. TDS affect the electrical conductivity of water.

(iv) *Fluorides.* Ground water is likely to have excessive concentrations if fluoride bearing rock formations exist. Industrial effluents also contribute fluoride compounds to water sources. Dental caries (decay of tooth enamel) in children due to drinking fluoride free water and fluorosis (mottling of teeth and bone/skeletal softness) due to excessive fluorides (more than 1.5 mg/L) occur when people use water over a long period.

(v) *Chlorides.* Chlorides of Ca, Mg and Na are present commonly in natural waters as they are highly soluble. Excessive concentrations may be due to sea water intrusion or contamination from domestic sewage or industrial wastes.

(vi) *Sulphates.* Calcium, magnesium and sodium sulphates are generally found to be present in water. $MgSO_4$ produces laxative effects and hardness.

(vii) *Nitrates.* Decomposed organic matter (from sewage) or agricultural (from fertiliser) drainage may join water sources and result in excessive nitrates. Infant mathemoglobeanemia (blue baby symptom, i.e. lack of oxygen in blood) is associated with excessive nitrates in drinking water.

(viii) *Phosphates.* These are nutrients. Excessive concentration of phosphates and the presence of nitrates and sulphates cause algal growth in lakes and reservoirs.

(ix) *Toxic chemicals.* Chemicals containing Zn, Cu, Pb, As, Ba, Cd, Cr,

Hg, Ni, Se, Th, cause damage to internal organs of the human body (kidney, liver, brain). Various organic compounds, pesticides, carbon tetra chloride (CCl_4), Vinyl chloride, Cyanides, Phenol, chlorinated hydrocarbons, etc. are highly toxic. Presence of iron in water leaves stains on clothes, imparts bad tastes and is responsible for bacterial growth.

(x) *Dissolved oxygen (D.O.).* Oxygen is absorbed by water from the atmosphere and natural water has a D.O. content of 8–10 mg/L depending on its temperature. Dissolved oxygen makes the water taste good. A minimum of 4–5 mg/L is essential for fish and aquatic life. Surface water will have more D.O. due to greater surface area in contact with atmosphere. Organic wastes utilise dissolved oxygen for the biological decomposition. Inorganic materials also utilise oxygen in reduced state for oxidation. So depletion of D.O. indicates utilisation by organic and inorganic compounds, i.e. contamination from domestic and industrial wastes.

Bacteriological

Water borne diseases are caused by pathogenic microorganisms, like viruses, bacteria, protozoa and helminthic worms. They remain in the intestines of sick human beings and when discharged through faecal/oral route, contaminate water sources.

Common diseases, caused by contaminated water are: gastroenteritis, typhoid, cholera, bacillary dysentery, amoebic dysentery, polio myelitis, infective hepatitis (jaundice), etc. Bacteria multiply at a very rapid rate. There is every possibility of spreading of water borne diseases in the community affecting human populations in no time. So faecal coliforms in drinking water should be removed completely.

It is difficult and time consuming to test for the presence of specific disease causing bacteria. So non-pathogenic (coliform group) bacteria is measured to reliably indicate water contamination from faecal matter (domestic sewage) as they abundantly exist in the intestines of human beings. Disinfection of water has eliminated water borne diseases in the society to a large extent.

3.1.4 Water Quality at Source

No water is free from possible pollution from domestic or industrial wastes as they are discharged into ground water, rivers and lakes. Adequate measures must always be taken against bacterial contamination at the source of water supply.

1. Ground water contains less amount of suspended impurities due to the presence of different layers of soil, which act as filters. Dissolved salts are more likely, as water is in contact with soil and rock formations. Dissolved oxygen is less when compared to surface waters. Presence of Fe, Ca, Mg, fluorides, chlorides (and salt water intrusion in coastal areas) in ground water is normal. Leachates from land fill, dissolved salts from liquid wastes can sometimes percolate into ground water.

2. In lakes and reservoirs, impurities settle down due to less flow, and absence of turbulent conditions. Re-oxygeneration from the atmosphere also occurs slowly. Through contamination, nutrients (N, P) enter into lakes causing more algal growth and more microbic aquatic life. Eutrophication adds to turbidity, organic load and stratification of layers from top to bottom, which cannot be easily reverted within a reasonable time. Hence, care must be taken to limit plant nutrients entering into lakes and to remove weeds and algal growth periodically. Water from lakes will be acceptable for drinking purposes, only if contamination is brought down to the permissible level.
3. River water contains more dissolved oxygen and is more palatable. Suspended solids and turbidity will be more. River waters are more susceptible to surface contamination.

Water pollution

If water contains impurities beyond a permissible level, it has harmful effects on the health of the users, which can be listed as follows:

(a) Bacteria, virus, or fluorides, nitrates, sulphates, etc., present in drinking water, cause diseases of serious nature.

(b) Organic matter (from wastes) present in water, consumes the dissolved oxygen for biological decomposition, thus rendering it unfit for aquatic life.

(c) When toxic chemicals from industrial effluents join the water sources, water becomes unfit for domestic use as well as for agriculture.

Presence of chlorides, nitrates and lack of dissolved oxygen are the main indicators of pollution. Polluted waters affect biodiversity. Certain aquatic life shows tolerance towards different degrees of pollution. Algae, diatoms, protozoa, rotifers, crustacea, and some varieties of fish are sometimes taken as biological indicators for the levels of surface water pollution.

3.1.5 Drinking Water Quality Standards

Drinking water must satisfy the following requirements:

1. Safety from water borne or associated diseases.
2. Wholesomeness, which means reasonable quantities of required chemicals (salts) should be present in water (distilled water is not ideal for drinking). Water should be clear, palatable and fresh and free from excessive concentration of chemicals.
3. Water must be free from all toxic chemicals that are injurious to life.

Based upon these requirements, the following permissible (standard) limits have been prescribed for potable drinking water by various agencies like BIS and

WHO. It should be noted that, two to three litres of water is required per person per day for drinking purpose, whereas the total requirement for all domestic uses put together is around 150–200 litre/head/day. But the total quantity of water is treated and supplied to each and every customer through a single pipeline in a town as distribution networks for separate pipe lines cost more. The parameters which need to be followed for drinking water are given in the following chart.

Drinking Water Standards

Parameters	Desirable maximum limits (mg/L) (except colour, turbidity and pH)
1. Colour (Hazen's scale)	5–10
2. Turbidity (NTU)	10
3. Taste and odour	Acceptable
4. pH	6.5 to 8.5
5. TDS	1000–1500
6. Hardness ($CaCO_3$)	200–250
7. Chlorides	250
8. Sulphates	150
9. Fluorides	0.8 to 1.5
10. Nitrates	45
11. Fe	0.1
12. Zn	5.0
13. Hg and Phenol	0.001 (each)
14. Cr, CN, As, Cu, Mn, Se, Pb, Cd	0.01 to 0.1 (each)
15. Coliform count (MPN)	Zero
16. Radio activity in terms of alpha and beta activities	Nil
17. Alkyl benzene sulphonate (detergents)	less than 1
18. Polynuclear aromatic hydrocarbons	0.001

3.1.6 Water Quality for other Purposes

Bureau of Indian Standards has prescribed standards of quality of water required in different industries. Water suitable for one industry may not be suitable for another industry.

For irrigation purposes, water should have less than 30 mg/L of BOD, 30 mg/L of suspended solids (S.S.) and less MPN value of faecal coliforms.

For Textile industry, Fe—0.25, TDS—50, S.S.—500, hardness—100, pH 6–8 are desirable upper limits.

For Paper industry, Fe—0.25–10, TDS—500, chloride—200, hardness 100–200, pH—5–9, and colour—5 are desirable.

For Boilers, turbidity—10 JTU, total hardness—40, total solids about 1000 are desirable.

For ice manufacturing, turbidity—5, hardness—50, alkalinity—40, TDS—300, Fe—0.2 are acceptable limits.

For tanning, Turbidity—20, Hardness—100, alkalinity—100, pH—8.

(Units of all parameters are as given in the drinking water standards)

All other properties should conform to the standards of drinking water quality. Industries may have their own purification units and may not depend upon the municipal water treatment systems.

3.2 WATER SUPPLY

Water demand and water distribution are the two basic aspects in any water supply system.

3.2.1 Per Capita Demand

Water is used for various purposes and the rate of consumption depends upon the purpose

(a) Domestic use (drinking, cooking, cleaning, washing, bathing, flushing toilets, etc.) requires about 100 L/head/day.

(b) Commercial places (hotels, cinema halls, bakeries, parlours) use about 45 L/head/day.

(c) Public places (schools, parks, hospitals, markets, railway stations, bus stands, etc.) need about 30 L/head/day. For hotels, hospitals, bakeries etc., individual requirements may be more than 40 L/head/day. But as they are location specific, average values are only taken while considering their per capita demand.

(d) Fire fighting needs substantial amount of water as a reserve against any contingency.

(e) Losses and wastage may account for about 5–10 L/head/day.

The total per capita demand (average) is about 200 L/head/day, which may vary from 100–300 L/head/day, depending upon factors like climate, duration of water supply, standard of living, drainage system, method and extent of charging the consumer, and pressure available in distribution mains. Water consumption is more in the morning hours (5–9 a.m.) and evening hours (5–8 p.m.). It varies hourly, daily and seasonally. This fluctuation in average rate of demand (Fig. 3.4) has to be taken care of in designing water carrying mains (in peak times 1.5–3 times more than average flow). Water supply projects are designed so as to cater to the community needs for a (design) period of 15–30 years taking into consideration future population growth, life span of pipes and appurtenances and financial considerations.

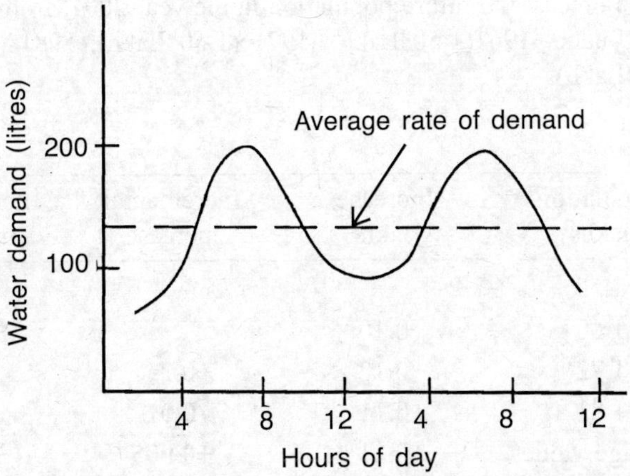

Fig. 3.4 Fluctuations in average rate of demand.

While designing a water supply system, a factor of safety of 1.5 times the average flow in purification plant design and upto 3 or 4 in the distribution system are adopted.

Population forecasts

(i) *Arithmetical increase method.* Population increases at a constant rate, i.e. average of increase in the previous decades is considered.

(ii) *Geometrical increase method.* Percentage growth is assumed to be constant i.e. per cent increase in population for each decade is calculated and the average of per cent increase in the previous decades is considered.

(iii) *Incremental increase method.* Increase in population is calculated as in the arithmetical increase method. But the increase cannot be expected to be the same in all decades. Hence, an extra element of increase (or decrease) in population over two decades is calculated to get the incremental increase value.

(iv) *Graphical method.* A graph is drawn between population vs time from the available data. This plot can be extended to obtain future population for any time period.

(v) *Method of comparison.* Towns having similar growth development pattern are likely to have similar trends in population growth. Future population of a town can be predicted by comparing with some other town already developed on the same basis.

Proper method of prediction is considered depending upon the development patterns of a town. If an average growth of a town is expected, arithmetical increase method is selected for predicting future population. If the town is going to follow a rapid development pattern, then geometrical method is adopted.

PROBLEM 3 Forecast the future population in the year 2001, from the figures of the previous years—1961(1.50 Lakh), 1971 (1.60 Lakh), 1981 (1.72 Lakh) and 1991 (1.83 Lakh).

Solution

Year	Population (lakhs)	Increase (lakh)	Incremental increase	Percentage increase
1961	1.50			
1971	1.60	+0.10		+6.7
1981	1.72	+0.12	+0.2	+7.5
1991	1.83	+0.11	–0.01	+6.8
	Average values	+0.11	+0.005	+7.0

(i) By arithmetical method, population in 2001 = population in 1991 + average increase of population over 3 decades 1.83 + 0.11 = 1.94 lakhs

(ii) By geometrical method, population in 2001 = population in 1991 + average percentage of increase in population over 3 decades × population

$$\text{in } 1991 = 1.83 + \frac{7\,(1.83)}{100} = 1.9581 \text{ lakhs}$$

(iii) By incremental increase method, population in 2001 = population in 1991 + average increase in population over 3 decades + average incremental increase in population over the two phases = 1.83 + 0.11 + 0.005 = 1.945 lakhs

(iv) By graphical method, as given in Fig. 3.5, population in 2001 = 1.95 lakhs

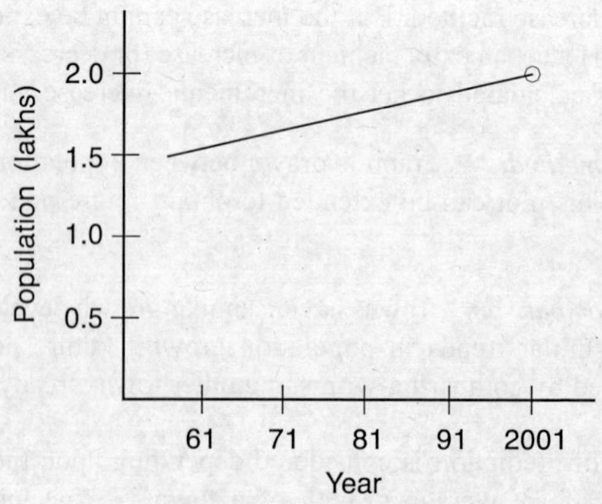

Fig. 3.5 Population forecasts (Graphical method).

PROBLEM 4 Calculate the quantity of water required for fire demand for catering to the needs of 100,000 population.

Solution

(i) Kuchling's formula

$$Q = 3182\sqrt{P},$$

where

P = population in thousands,

Q = water quantity required in litres per minute (L/min) for extinguishing fire

$$Q = 3182\sqrt{100} = 31{,}820 \text{ L/min.}$$

(ii) Freeman's Formula

$$Q = 1136\left[\frac{P}{5} + 10\right] = 34{,}080 \text{ L/min.}$$

(iii) National Board of Fire under writer's formula

$$Q = 4637\sqrt{P}\,[1 - 0.01\sqrt{P}] = 41{,}733 \text{ L/min.}$$

Note: 1. About 1 to 5 L/head/day may be taken as fire demand. It depends also on factors like, possibility of fire occurrence, nature of the township.

2. Water should be made available in storage reservoirs and fire hydrants must be suitably provided for extinguishing fire.

3.2.2 Water Collection from Source

(i) *Intakes.* Water enters into an intake well and is pumped to a purification plant. Intake wells are located on the upstream side of the river such that maximum quantity of good quality water can be drawn even during summer periods. A strainer (screen) is needed for eliminating suspended and floating impurities. Multiple inlets are provided for taking water from different levels (Fig. 3.6). Water can be drawn, using similar intake devices from canals and reservoirs.

(ii) *Tube wells.* The main components are a solid casing, a pipe column, a pump head (attached at the top end) connected to a motor and a screen connected to the suction (lower) end (Fig. 3.7). After drilling operation, casing is set in position and sealed.

(iii) *Infiltration wells and galleries.* Infiltration pipes and galleries (horizontal wells) are located below ground level, on the banks of a river bed. Water infiltrates into the net work, collected and pumped for storage and use. They are packed with gravel to prevent the entrance of sand (Fig. 3.8).

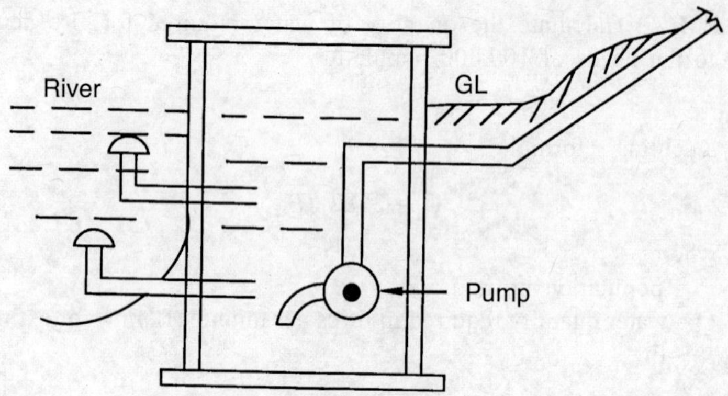

Fig. 3.6 Intake.

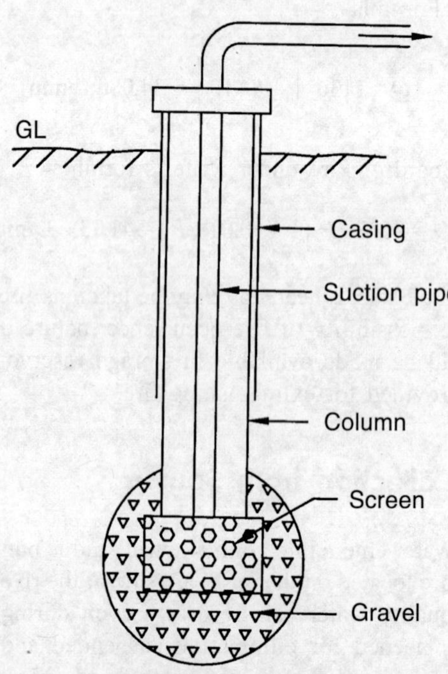

Fig. 3.7 Tube well.

3.2.3 Pipes and Appurtenances

Water is carried through a network of pipes. Different materials are used for the pipes and small lengths are joined together. A number of valves and other appurtenances that are used are explained here.

Pipe Material

(i) Cast iron or steel pipes (of large diameters of 500–900 mm) are used for the distribution and conveying mains, which require strength and durability.

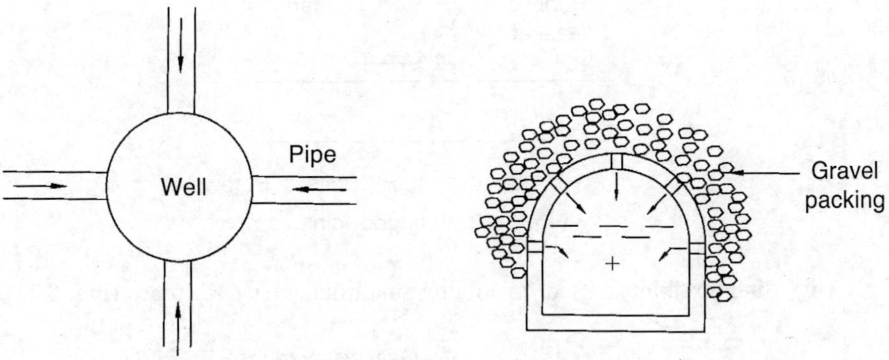

Fig. 3.8 Infiltration pipes, wells & galleries.

Advantages that the steel pipes have are: long life, corrosion resistance, easy jointing facilities and moderate costs.

(ii) R.C.C. pipes have bigger diameters and P.C.C. pipes have smaller diameters (less than 500 mm). They can resist external loads and are easy to construct. They do not get corroded.

(iii) Asbestos cement (A.C.) pipes (less than 600 mm dia) are used in water distribution. They are light and easily assembled.

(iv) Plastic LDPE (60 mm dia.), HDPE (300 mm dia.) pipes or rigid PVC pipes (15 to 150 mm dia.) are used. They are not durable. They are light and used for making connections for water supply in houses.

Pipe Joints

Pipes are manufactured in small lengths for convenience in handling and transporting operations. These small pieces of 2 to 6 m lengths are joined together after placing them in proper position.

(i) Spigot & socket joint is mostly used for cast iron pipes. Yarn of hemp is tightly wrapped around spigot end of the pipe and then molten lead is placed to fill the space between the socket and spigot ends (Fig. 3.9).

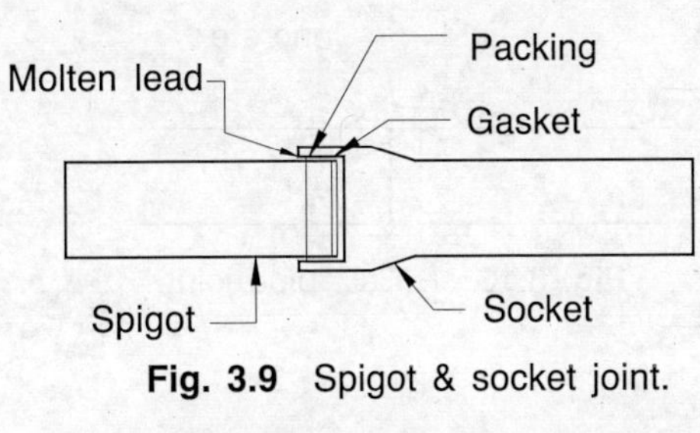

Fig. 3.9 Spigot & socket joint.

(ii) Flanged joint is used for temporary pipe lines, as they can be reassembled with ease. A piece of hard rubber washer is placed in between the two flanges. The joint is made water-tight by using screws (Fig. 3.10).

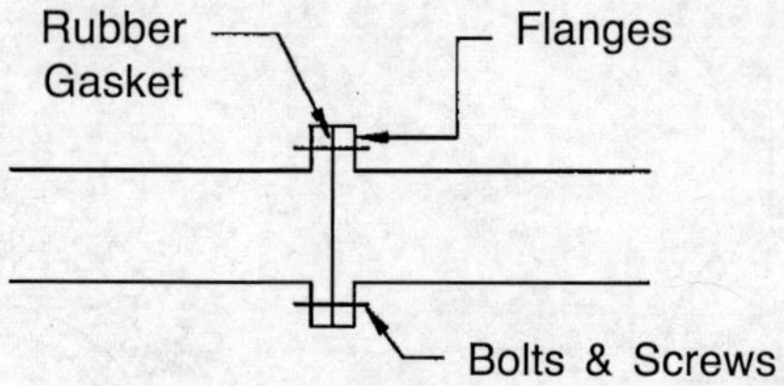

Fig. 3.10 Flanged joint.

(iii) Simplex joint is used for joining small diameter A.C. pipes (Fig. 3.11).

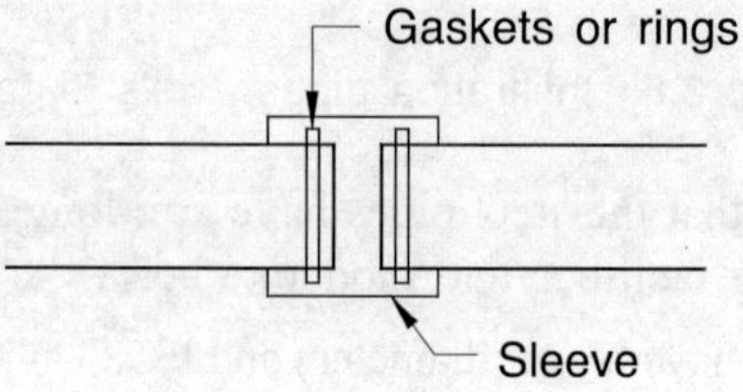

Fig. 3.11 Simplex joint.

(iv) In an expansion joint, the socket is allowed to freely slide on the spigot end of the pipe, to take care of the effects of temperature on the pipe material (Fig. 3.12).

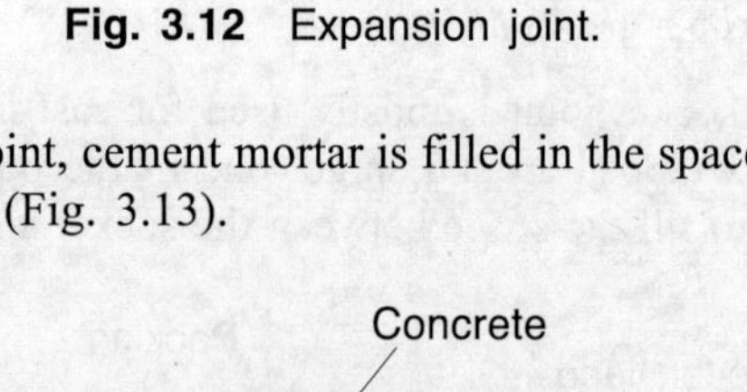

Fig. 3.12 Expansion joint.

(v) In RCC pipe joint, cement mortar is filled in the space between the pipes and the collar (Fig. 3.13).

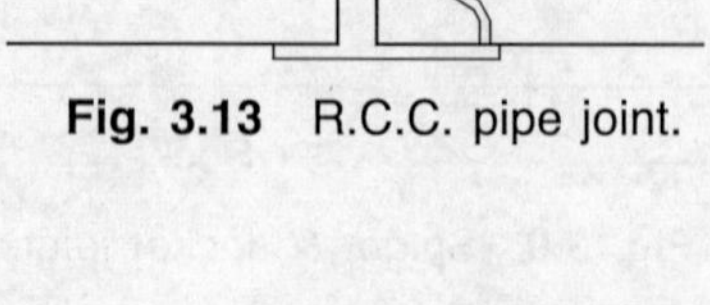

Fig. 3.13 R.C.C. pipe joint.

Valves

(i) Air relief valves are provided to remove accumulated air at elevated or higher (summit) points in flow through pipes (Fig. 3.14). Air gets

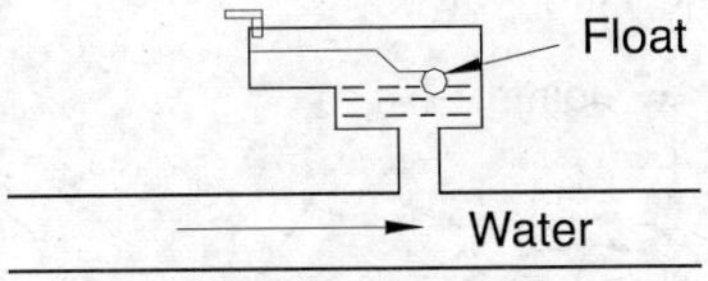

Fig. 3.14 Air relief valve.

accumulated in the chamber and water level is depressed. The float sinks down and thus opening the relief valve for the escape of air.

(ii) Safety valve or pressure relief valve protects pipes from excessive pressures of flow (Fig. 3.15). The pressure is released when some portion of water escapes out by opening the valve.

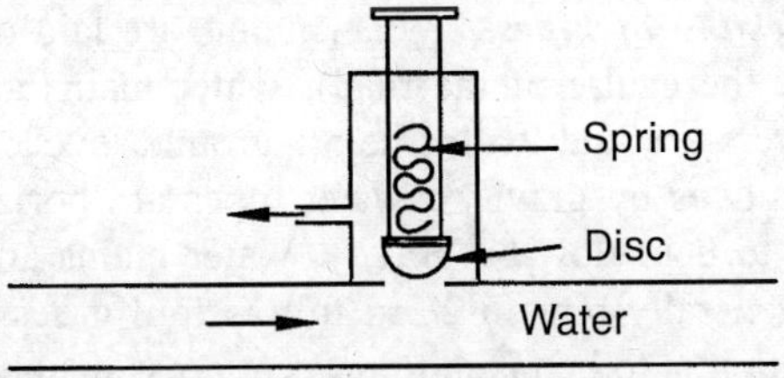

Fig. 3.15 Safety valve.

(iii) Reflux (check) valve allows flow in one direction only (Fig. 3.16). When installed on the delivery side of the pump set, back flow is prevented.

Fig. 3.16 Reflux (check) valve.

(iv) Sluice valve (to regulate flow of water) (Fig. 3.17) and gate valves are used to control (stop) the flow during maintenance and repair.

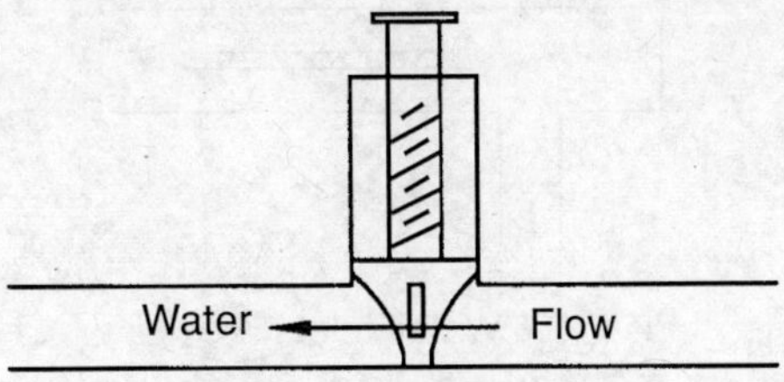

Fig. 3.17 Sluice valve.

(v) Drain (scour) valves are provided at low points to remove silt and other deposits.

(vi) Service connection is the link between water supply on the main road and the pipe line in the consumer's premises (Fig. 3.18).

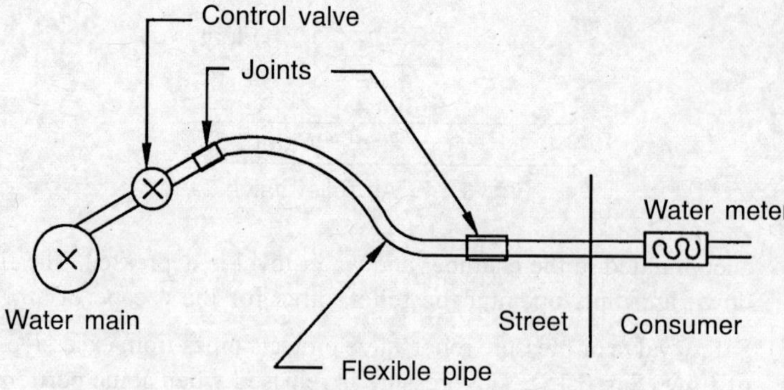

Fig. 3.18 Service connection.

Laying and Testing of Pipes. Water mains are laid on one side of a road (sewers are laid at the centre of the road). Water mains are about 1 m. below ground level. (Sewers are laid 2–4 m below ground level). Water moves with a pressure (sewage moves by gravity). Water pipes are horizontally laid whereas sewers have slope to help flow by gravity. Water mains are always full, due to pressure. Sewers generally remain 2/3rd full as foul gases are released.

Utmost precaution is to be taken to prevent cross connections between water mains and sewers. Leakages in water mains causes not only wastage of water but gives opportunity for sewage to enter water mains. After laying pipes, they are jointed and tested for withstanding pressures and leakage.

Pumps

(i) Reciprocating pumps. In the forward stroke of a plunger, water enters from a suction pipe and in the return stroke, water moves into the delivery pipe (Fig. 3.19).

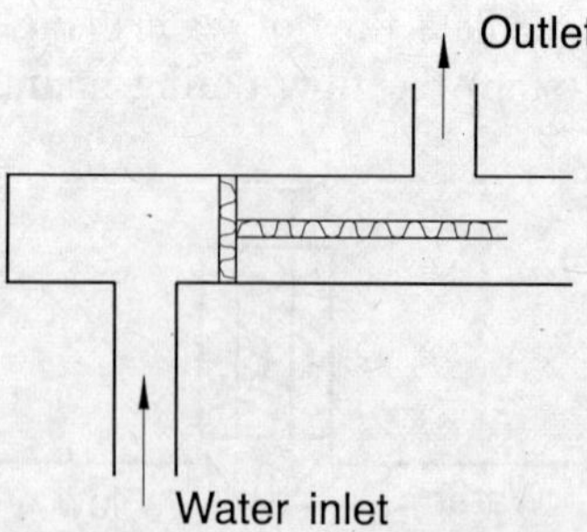

Fig. 3.19 Reciprocating pump.

(ii) Air lift pumps. Compressed air is forced into the well. The mixture of air and water rises through educater pipe and moves upward (Fig. 3.20).

(iii) Centrifugal pumps. In volute type, impeller vanes discharge water into a gradually expanding spiral case. Velocity of flow remains same at all

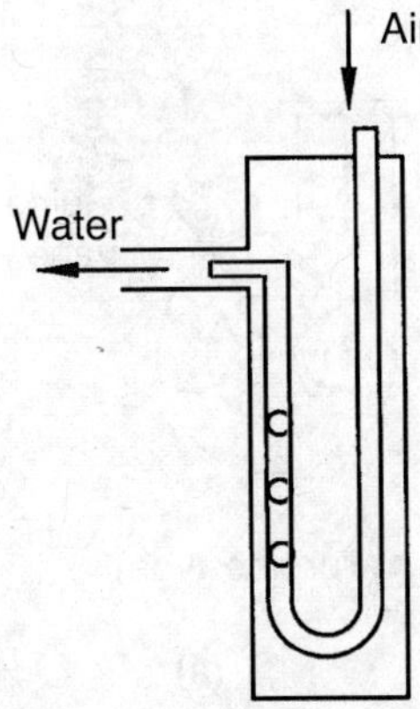

Fig. 3.20 Air lift pump.

points in the casing. In turbine type, impeller is surrounded by stationary guide vanes. Velocity head changes into pressure head due to diffuser vanes (Fig. 3.21(a)).

(iv) In a rotary pump, gears rotating in opposite directions push the water upward.

A typical installation is shown in Fig. 3.21(b).

PROBLEM 5 Determine the Horse power of a pump given that:

Specific weight of water (W) = 1000 kg/m^3

Total head to be developed by pump (H) = 40 m (including length of suction pipe + loss head in the system)

Efficiency of the system (η) = 80% = 0.8

Discharge of the pump (Q) = 0.0075 m^3/s

Solution

$$\text{H.P.} = \frac{WQH}{75\eta} = \frac{1000 \times 0.0075 \times 40}{75 \times 0.80} = 5 \text{ H.P.}$$

(diameter of the economical main to balance pipe costs and pumping costs = 1.0 to 1.22 $\sqrt{Q} = 1 \times \sqrt{0.0075}$ = 0.86 m)

3.2.4 Water Distribution

The purified water should be supplied to a consumer, with adequate pressure (0.6 to 1.5 kg/cm^2) and in required quantity through an efficient distribution system. Pipes, valves and other appurtenances are widely used in water distribution system and are costly. Cost increases exponentially with increase in diameter. So,

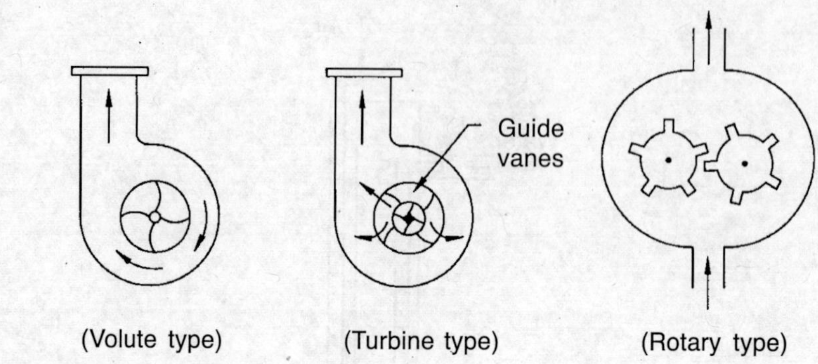

(a)

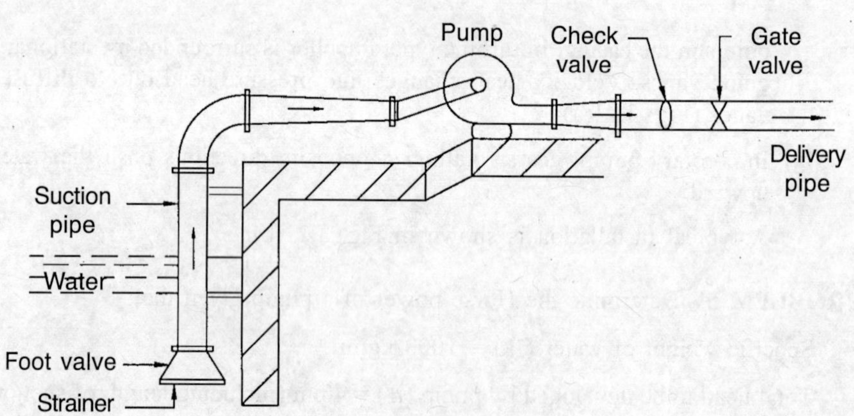

(b)

Fig. 3.21 (a) Centrifugal pumps, (b) Centrifugal pump installation.

economy is a significant factor in making the designs of pipes. Distribution system should also cater to the needs of emergency situation such as fire fighting which requires uninterrupted supply of water. The following are the basic requirements of a distribution system:

1. To supply water with adequate pressure (15 m or 150 kPa)
2. To meet the water demand (including fire needs)
3. To reduce capital construction costs by choosing proper pipe diameters and all appurtenances
4. Easy operation and maintenance
5. To reduce the losses and cross connections in the system

In any distribution method, water flows from source to treatment plant and from treatment plant to distribution area. This may be due to (a) gravity (depending upon ground elevation), (b) pumping or (c) combination of the two, as shown in Fig. 3.22. Storing water in service reservoir for distributing the same at the scheduled hours is more reliable.

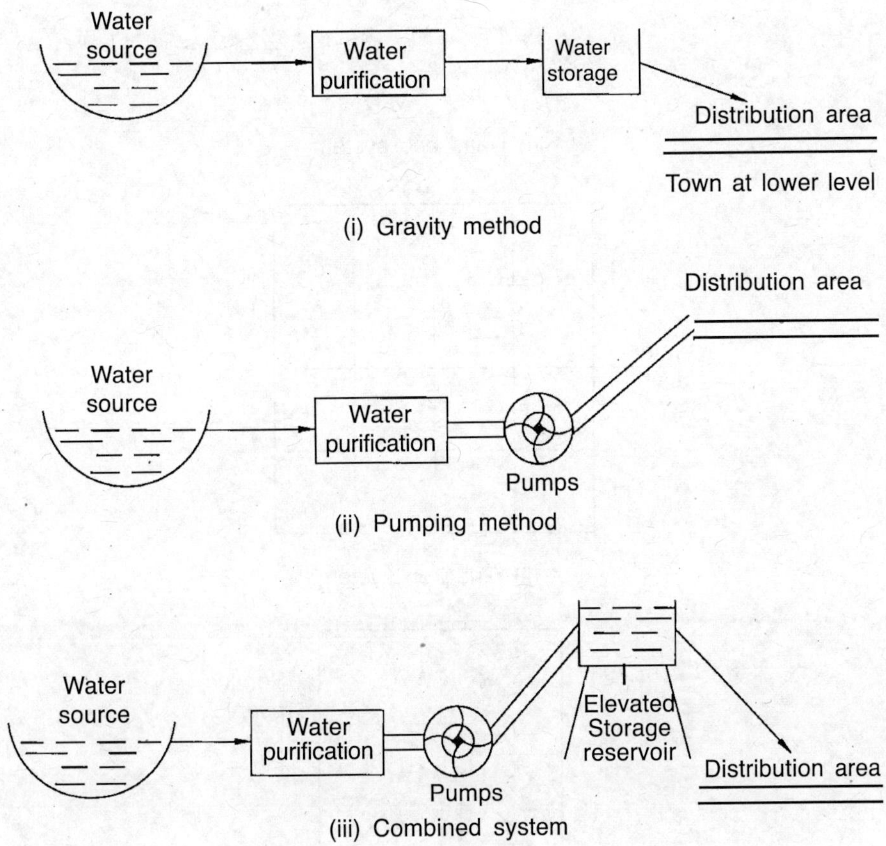

Fig. 3.22 Water distribution methods.

Layouts

Water flows from distribution reservoir to the consumer by different patterns depending upon the layout of roads and distribution of houses in a colony. In a dead end system (Fig. 3.23(i)), water mains are not interconnected. They branch off like a tree. If houses are scattered, it is economical to have this type of distribution layout. In a grid iron system (Fig. 3.23(ii)), mains are interconnected to keep water under circulation and flow in different directions. For colonies with a rectangular road layout pattern, this system is reliable.

In a radial system (Fig. 3.23(iii)), service reservoir is placed at the centre from where water flows radially (along the road direction). It may be similar to a dead end system if there is no interconnection among water mains. In a circular system, water moves on a periphery and is interconnected as in a grid iron system.

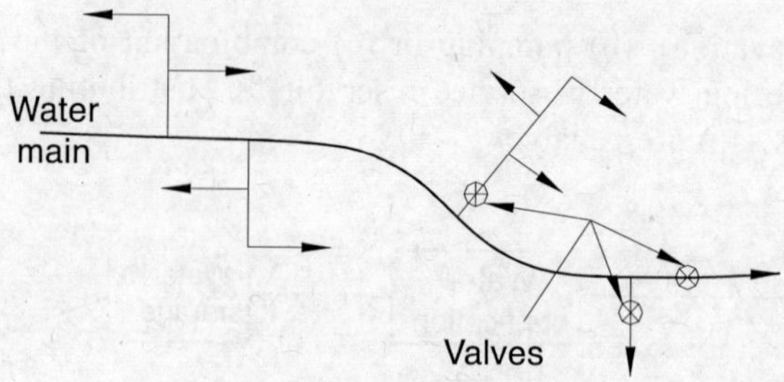

(i) Dead end system

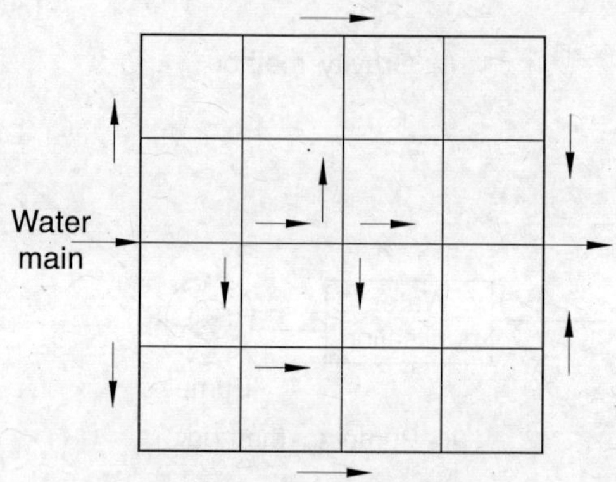

(ii) Grid iron system

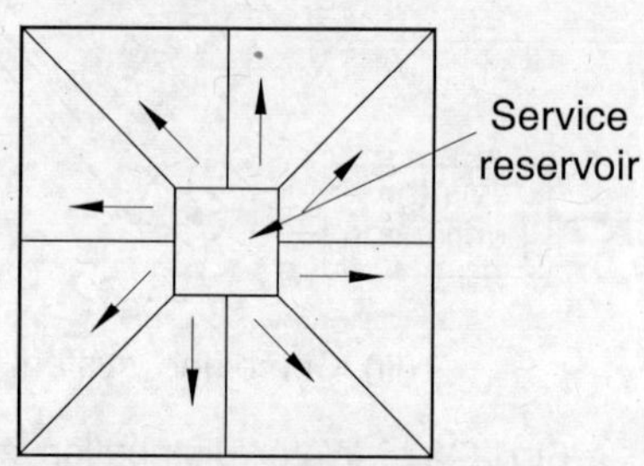

(iii) Radial system

Fig. 3.23 Distribution systems.

Comparison. (i) Tree system has less appurtenances but during repair, water has to be stopped in a large number of areas.

(ii) Grid iron system has more interconnections and valves. Construction costs are high and design work needs expertise.

3.2.5 Service Reservoir

Water may be directly pumped into the mains to reach the consumer, or water may flow from a service (elevated) reservoir (Fig. 3.24) by gravity. The reservoir is a balancing tank. Extra water is stored and used whenever

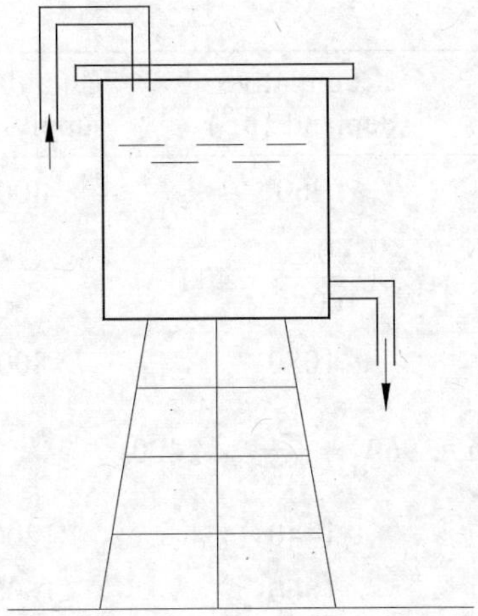

Fig. 3.24 Service reservoir.

required. When water demand exceeds the quantity pumped directly, more water can be drawn from reservoirs. Capacity of a service reservoir is obtained using a mass curve.

A graph is drawn for cumulative flow with time for water demand and pumping capacity as in Fig. 3.25. The vertical intercept (showing the difference between demand and pumping) is taken as the capacity of reservoir.

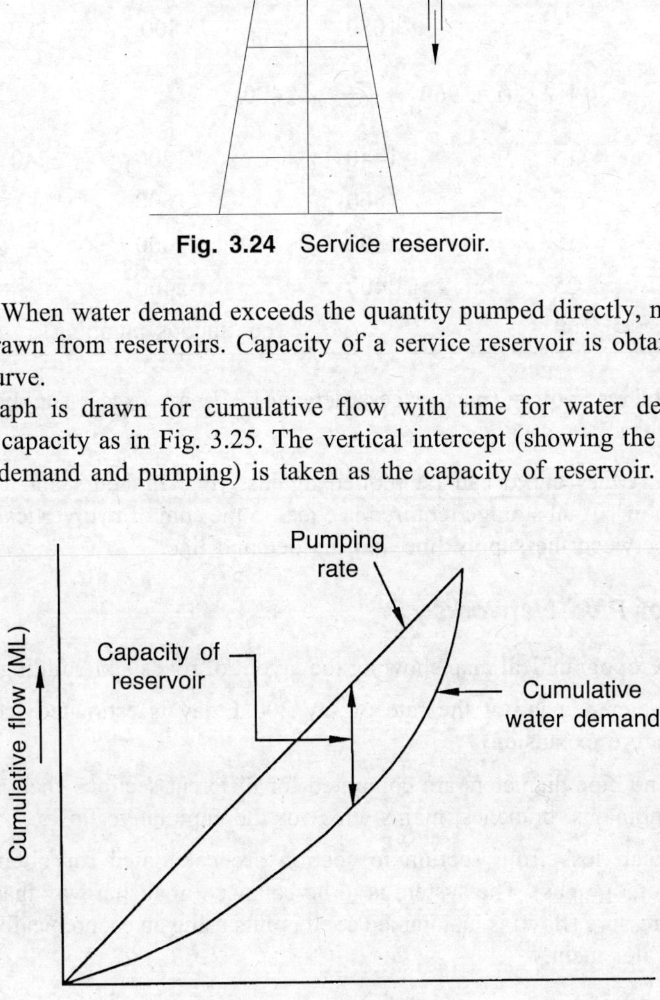

Fig. 3.25 Mass curve.

PROBLEM 6 Determine the capacity of a reservoir when the demand is 2400 m^3/day and the rate of pumping is 100 m^3/hour, when the per cent demand in different time intervals is given.

Solution

Time	Per cent demand	Cumulative demand (m^3)	Cumulative supply (m^3)	Excess demand	Excess supply
6–10 a.m.	40	960 (i.e. $\frac{40}{100} \times 2400$)	400	560	
10–2 p.m.	5	1080 (i.e. $960 + \frac{5}{100} \times 2400$)	800	280	
2–6 p.m.	15	1440	1200	240	
6–10 p.m.	5	1560	1600	40	
10–2 a.m.	10	1800	2000	–	200
2–6 a.m.	25	2400	2400	–	–
Total	100		(continuous pumping)		

Capacity of reservoir = (max. excess demand) + (max. excess supply) = 560 + 200 = 760 m^3

Note: Mass curve can be plotted against time in hours and cumulative demand in m^3. Total storage required is equal to the sum of two vertical ordinates (or one) between the supply line and the demand line.

Design of Pipe Networks

(i) A topographical map showing the layout of roads and buildings is taken.

(ii) Water demand at the rate of say 200 L/day is estimated (considering future expansion).

(iii) The pipe diameters are computed for different sections such as laterals, submains, branches, mains, etc., for the cumulative flows.

(iv) Head loss from section to section are calculated (or obtained from nomographs). The system is to be balanced in such a way that adequate pressure (flow) is maintained at all points using an economically designed water main.

The following formulae are useful.

- Hazen William's Formula:

$$V = 0.85\ C\ r^{0.63}\ s^{0.54}$$

where

V = velocity of flow (m/s)

C = a constant depending upon the pipe material and roughness (120–130)
r = hydraulic mean depth (m) (area/wetted perimeter)
s = loss of head (slope) per m length.

- Darcy Weishbach equation:

$$h = \frac{fLV^2}{2gd} = \frac{fL}{2gd}\left(\frac{Q}{A}\right)^2$$

where
h = loss of head (m)
f = friction factor (0.02 to 0.005)
L = length of pipe (m)
V = velocity of flow (m/s)
g = acceleration due to gravity (9.81 m/s^2)
Q = discharge (m^3/s) = A.V
A = area of cross section (m^2), or $\left(\frac{\pi d^2}{4}\right)$
d = pipe diameter (m)

- Manning's Formula:

$$V = \frac{1}{n} r^{2/3} s^{1/2}$$

where
n = Manning's rugosity coefficient (0.012–0.014)

PROBLEM 7 Find the diameter of a water main given that,
Length of the main, L = 1000 m
Frictional coefficient, f = 0.005
Flow through pipe, Q = 0.5 m^3/s
Height of the reservoir, h = 40 m
Minimum pressure in the distribution system = 10 m.

Solution

$$h = (40 - 10) = \frac{fL}{2\,gd}\left[\frac{Q}{\pi d^2/4}\right]^2$$

$$= \frac{0.005 \times 1000}{2 \times 0.98 \times d}\left(\frac{0.5 \times 0.5 \times 4 \times 4}{3.14 \times 3.14 \times d^4}\right)$$

d = 0.5 m.

Equivalent pipe method

Equivalent pipe is an assumed (imaginary) main to replace a system for equal head loss under same quantity of flow.

AC is an equivalent pipe to replace the system of pipes, e.g. *ABC* and *ADC*

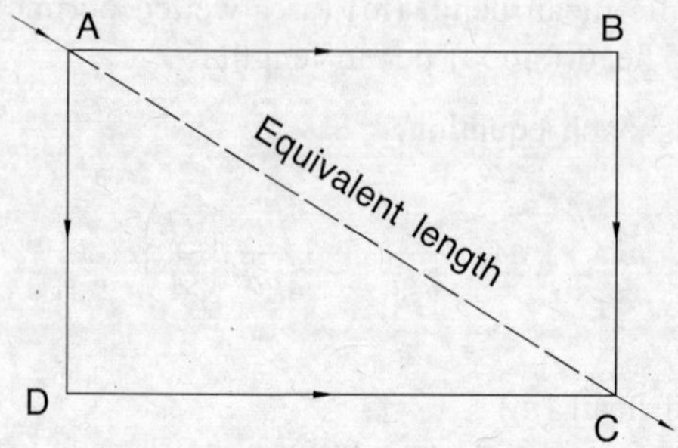

Fig. 3.26 Equivalent pipe.

in Fig. 3.26.

Head loss from *A* to *C* is calculated from two routes *ABC* and *ADC*. For that head loss and same discharge, the (equivalent) length of *AC* main is calculated.

Hardy cross method

This method is to keep the head loss from one direction (*ABC*) approximately equal to that from the other direction (*ADC*). Selection of pipe diameter and quantity of flow through such system is economical and efficient. In a small loop *ABCD* (Fig. 3.27), discharges and lengths of the pipes along *ABC* and *ADC* are known. Pipe diameters are selected reasonably with experience. Loss of heads, (H_1) and (H_2) from two routes *ABC* and *ADC* are computed. If they are not the same, some adjustment is to be done for either pipe diameters or flows. If the difference is small, correction to discharge is sufficient. Otherwise pipe diameters are to be revised to balance head losses from two directions. The value of correction to the flow can be obtained using the formula

$$q = \frac{H_1 - H_2}{n\left\{\dfrac{H_1}{Q_1} + \dfrac{H_2}{Q_2}\right\}}$$

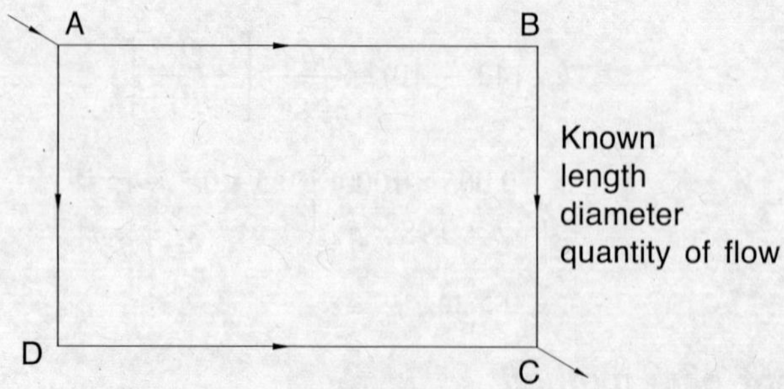

Fig. 3.27 Hardy-cross method.

where n = 1.85

That is, flow from the direction *ABC* may be increased by an amount equal to q and the flow from other direction may be reduced accordingly. Then again loss of head in each direction is calculated. The process is continued till they are same.

PROBLEM 8 Given that flow through the system, $Q = 50\ m^3/h$

AB has 400 m length and 40 cm diameter

BC has 300 m length and 30 cm diameter

AD has 300 m length and 35 cm diameter

DC has 300 m length and 30 cm diameter

Find the equivalent length of pipe *AC* to replace the network.

Solution

Pipe	Flow (m^3/h)	Length (m)	Diameter (cm)	Loss of head from Hazen Williams monogram (1 in 1000)	Loss of head for total length (m)
AB		400	40	0.10	0.04
BC		300	30	0.35	0.105
AD		300	35	0.25	0.075
DC		300	30	0.35	0.105

Loss of head in the direction of *ABC* = 0.145 m.
Loss of head in the direction of *ADC* = 0.180 m.
Total loss of head = 0.145 + 0.180 = 0.325
Replacing a single pipe of 30 cm dia, the

$$\text{equivalent length of pipe} = \frac{0.325}{0.35} \times 1000$$

i.e., 930 m instead of 1300 m. for the same head loss development

PROBLEM 9 In the above network, applying Hardy Cross method, balance the flows in two directions for obtaining same head loss in both the directions.

Solution

When Q is 50 m^3/h, the difference in loss of head in two directions *ADC* and *ABC* is 0.180 – 0.145 = 0.035 m.

$$q_1 \text{ (correction to flows)} = \frac{0.035}{1.85\left[\dfrac{0.325}{100}\right]} = 0.0006 \ \text{m}^3/\text{h}.$$

Increase the flow in the direction of *ABC* and decrease the flow in the direction of *ADC* by the same amount q_1 and determine the loss of head in the directions of *ABC* and *ADC*. Continue the process till the loss of head in two directions is almost same.

3.3 WATER PURIFICATION

3.3.1 Necessity of Purification

For domestic supply, it is necessary to make the water wholesome and pleasant to drink. Water contains suspended solids, turbidity, bacteria and a number of dissolved salts. They have to be removed so that water is fit for domestic use. To make water consumable, the following factors should be taken into consideration:

- Safety for preventing water borne diseases.
- Palatability (mineral content, temperature) from aesthetic considerations.
- Wholesomeness of water taking the dissolved salts into account (in desirable concentrations).
- Ensuring water quality so as to attain the permissible standards till the water reaches the consumer.
- Removing turbidity, colour, taste and odour which cause the water to be unacceptable.

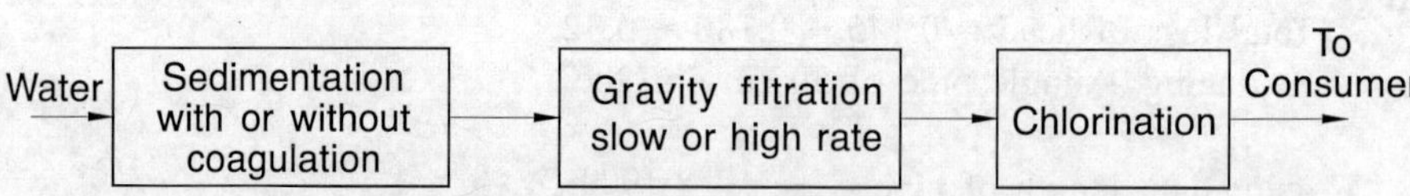

Fig. 3.28 Conventional treatment plant.

- Supplying adequate quantities of water for various domestic purposes.

A conventional treatment plant (Fig. 3.28) consists of three main units, viz. sedimentation (with or without coagulation), filtration and disinfection. Dissolved salts are removed only when necessary.

Provision of screens (at intake wells) exclude coarse material from source of collection. Plain sedimentation removes (about 35%) suspended solids and filtration removes more than 90% of suspended solids, turbidity and almost 95% bacteria. Coagulation process removes fine suspended and colloidal matter. Disinfection kills all the bacteria. Aeration helps in controlling the tastes and odours of water.

3.3.2 Sedimentation

Sedimentation is one of the most widely used processes in treatment of water, second only to perhaps chlorination. The simplest method of removing suspended impurities is by plain sedimentation. The water is allowed to stand quiescent or move very slowly through natural or artificial basins until the suspended impurities settle to the bottom and relatively clear water is obtained.

It is a process of getting the particles, heavier than water, settle due to gravitational forces under calm and quiescent conditions. Particles settle as discrete (single, unattached) particles in a plain sedimentation tank. Particles combine together (coalesce, flocculate) and settle quickly in coagulation—sedimentation tank. Sometimes sludge layer (blanket) settles as a single layer or gets compressed before settling.

In this process, two forces simultaneously act on a particle. A downward pull or a driving (impelling) force and a frictional resistance, i.e. drag force, acting upward on the particle. The resultant force decides whether a particle settles, floats or is carried away with flow.

The settling velocity of the particle depends upon (a) horizontal flow velocity of water, (b) shape and size of the particle, (c) relative density of the particle, and (d) temperature of the water; it is usually computed from Stoke's law which holds good only for particles of diameter less than 1 mm and Reynold number less than 1. In this range, viscous forces are large in relation to inertia. For particles of greater diameter and higher Reynold's number, drag becomes more prominent. When diameter of the particles increase above 10 mm and Reynold's number over 2000, Newton's law for frictional resistance or drag is applied. Most of the particles in water treatment settle well within the region of Stoke's law while sand grains and heavy particles settle in transition range, i.e. in between Stoke's and Newton's law.

Equating the above two forces, Stoke's law for settling velocity is derived:

$$(\rho_s - \rho_w)vg = C_D A \rho_w \frac{V_s^2}{2}$$

where ρ_s, ρ_w are density of particles and water respectively,

v = Volume of solid (spherical) particle

g = acceleration due to gravity

C_D = coefficient of drag = $\dfrac{24}{\text{Re}} = \dfrac{24\,\mu}{\rho_w V_s d}$

Re = Reynold's number

μ = viscosity of water

A = projected area of solid particle

d = particle diameter

s = sp.gr. of particle

T = water temperature

V_s = settling velocity of particle

Rearranging the equation, we have

$$V_s = \frac{g}{18}\frac{(\rho_s - \rho_w)\,d^2}{\mu} = \frac{g}{18\mu}(s-1)\,d^2 \quad \text{or} \quad 418\,(s-1)\,d^2\left(\frac{3T+70}{100}\right)$$

In a continuous flow sedimentation tank, settling efficiency depends upon (i) size, shape and specific gravity of particles, (ii) temperature, viscosity and density of water, (iii) water flow velocity, and (iv) disturbance from inlets and outlets and calm conditions.

Sedimentation basins are now operated almost exclusively, in which water flows through the basin continuously in a horizontal direction and leaving at the same rates. Horizontal flow tanks may be either rectangular or circular. Circular, horizontal flow tanks may have

(a) centre feed with radial flow

(b) peripheral feed with radial flow or

(c) peripheral feed with spiral flow.

However, in current practice, only the first is in common use. Continuous removal of settled sludge by revolving scrapers is preferred.

The ideal inlet (i) reduces entering velocity to prevent pronounced currents towards the outlet (ii) distributes the water uniformly over the cross section of the tanks and (iii) mixes it with water already in the tank to prevent the entering water (which is usually denser than the water already in the tank) from sinking to the bottom and moving along the bottom towards the outlet.

The outlet may be a free falling weir which skims the water from the top surface. Many rectangular tanks are now constructed with outlet channels and weirs extending across the ends and upto the sides of the tank for a distance of

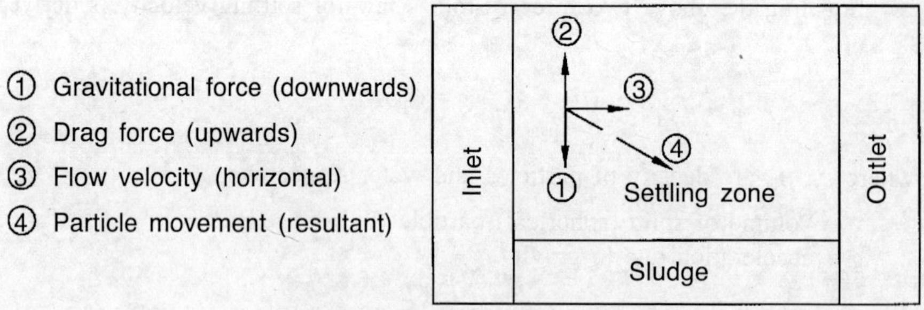

Fig. 3.29(a) Settling basin.

one-third or more of the length. Submerged weirs and submerged outlet parts attached to the end wall which discharge into a treated water conduits are also used.

A particle will have a horizontal velocity due to flow and vertical downward settling velocity. The resultant of these two components decides (in favour or not), the settling of a particle. An ideal settling basin is shown in Fig. 3.29(a).

Particle removal is not very efficient when water is not quiescent as:

(1) Water is in continuous flow and not as quiescent (still) as expected.
(2) Some portion of water flows through straight (just as a dye), the volumetric flow then does not utilise completely the volume of the tank provided as per detention time requirement.
(3) All particles are not of uniform characteristics.
(4) Eddies, short circuiting, disturbances due to inlet and outlet conditions reduce favourable conditions of sedimentation.

However, higher efficiencies are not required in plain sedimentation tanks.

3.3.3 Design Parameters

Detention time

It is the theoretical time for which the water is retained in the settling tank. It is taken as 2–4 hours, more for plain settling and less for coagulation units.

$$\text{Detention time} = \frac{\text{Volume of the tank}}{\text{Volumetric flow rate}} = \frac{\text{L.B.D.}}{\text{Q/T}} = \frac{\text{Length} \times \text{Breadth} \times \text{Depth}}{\text{Quantity of flow} \div \text{Time}}$$

Increase in the percentage of removal will be small even if the detention time

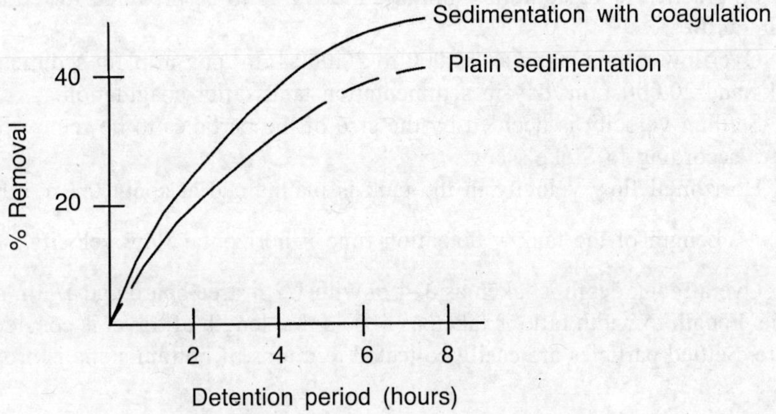

Fig. 3.29(b) Effect of detention period.

is increased beyond a certain stage as shown in Fig. 3.29(b). By increasing the detention time, volume of the tank will be more for the same flow. It is uneconomical and does not increase the efficiency considerably. Flow through period is the actual time taken for a dye to travel from inlet to outlet. This will be less than detention time. Efficiency of sedimentation depends upon the ratio of flow through period to detention time.

Overflow rate

Settling velocity decides whether a particle settles or not. Different particles (size, shape and characteristics) will have different settling velocities.

$$\text{Horizontal flow velocity} = \frac{\text{discharge}}{\text{cross sectional area}} = \frac{Q}{BD}.$$

A particle settles only if it strikes the depth (D) before moving the entire length (L) of the basin. Hence,

$$\text{Settling velocity of the particle} = \frac{D}{L} \times \frac{Q}{BD} = \frac{Q}{LB}$$

This means that, all particles having a settling velocity of more than (Q/LB) the overflow rate will settle in a sedimentation basin. However, the percentage of removal would vary depending upon the position of the particle and its length of travel or time required for travel in the tank.

This has given rise to

- Surface area loading is an important design parameter.
- Settling efficiency is independent of depth, as per the theory.

This means the efficiency of settling depends upon Q/A, known as overflow rate or surface area loading. It appears depth is not a very important factor for settling. But on practical considerations, minimum depth is to be provided for collection of impurities.

Overflow rate is taken as 12,000 to 20,000 L/m^2/day in plain sedimentation tanks and 20,000 L/m^2/day in sedimentation tanks after coagulation.

Settling velocity is decided by the size of the particles to be removed from water according to Stoke's law.

Horizontal flow velocity in the tank is maintained at about 0.3 m/minute.

$$\text{Length of the tank} = \text{detention time} \times \text{horizontal flow velocity}.$$

Overall tank depth is taken as 3–4 m with 1/2 m free board and 1/3 m sludge zone. Length to width ratio is taken as 4–5 so that length of travel is considerably more. Settled particles are easily collected to drains, if bottom slope is provided.

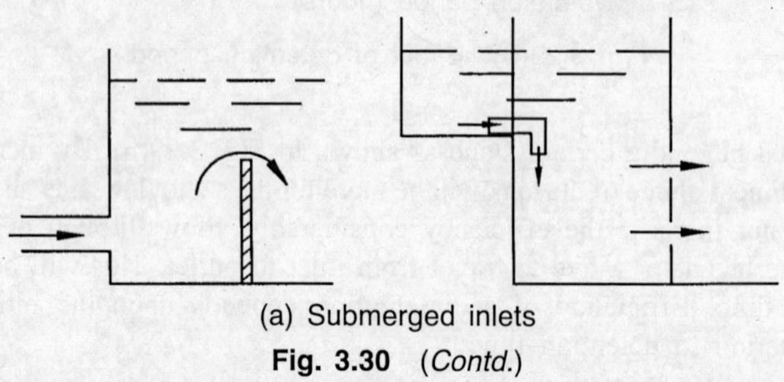

(a) Submerged inlets

Fig. 3.30 (*Contd.*)

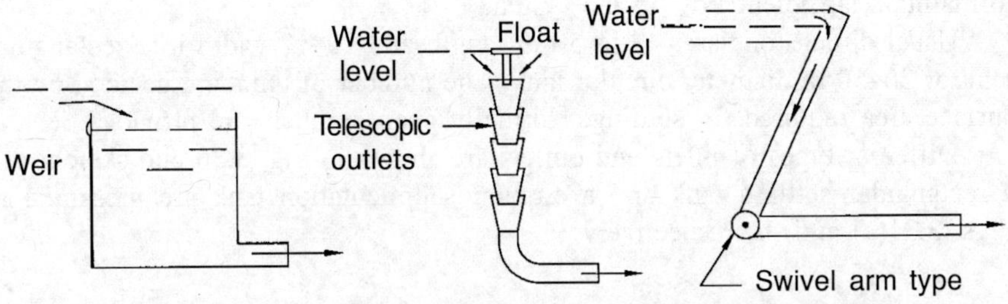

(b) Outlets

Fig. 3.30 Inlets and outlets.

Particle collection depends upon the cleaning device and cleaning interval.

Outlet weir loading is assumed to be 300–600 m^3/day/meter length of weir,

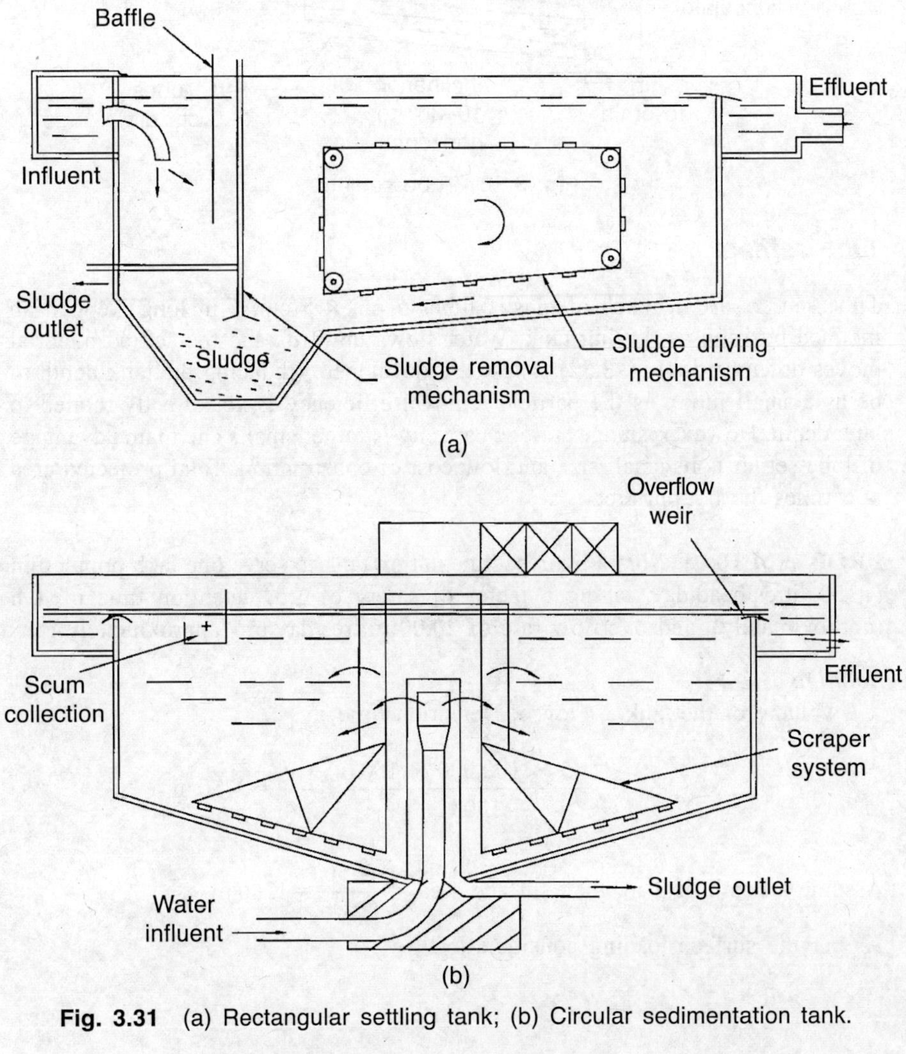

Fig. 3.31 (a) Rectangular settling tank; (b) Circular sedimentation tank.

to maintain sufficient flow in the system.

Usual dimensions are: 40 to 50 m length and 10–12 width in a rectangular tank or 30–50 m diameter circular tank. The number of units are decided by the surface area required for settling. Normally even numbers are provided.

Different types of inlets and outlets are shown in Fig. 3.30 and sketches of a rectangular settling tank and a circular sedimentation tank are presented in Figs. 3.31(a) and (b) respectively.

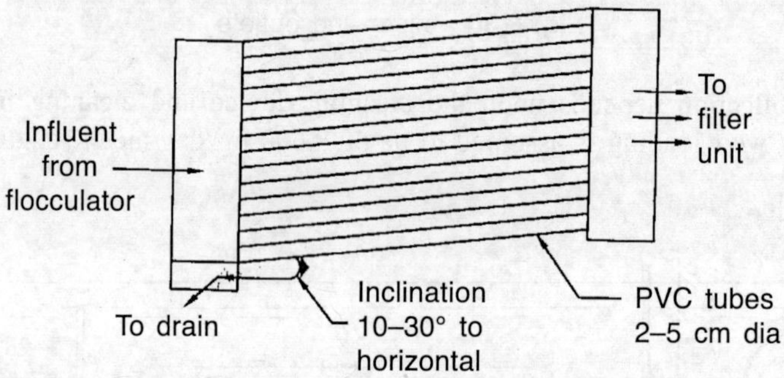

Fig. 3.32 Tube settlers.

Tube settlers

Tube settlers are a bunch of plastic tubes 5 cm × 5 cm, 1 m long, kept in an inclined position as settling tank. Water flows upwards and the settled material moves downwards (Fig. 3.32). These are developed, (reducing the tank depth to be as a small tube), as the particle removal efficiency is not directly related to tank depth. However, sludge storage capacity is rather small. The main advantage of tube settler is its small size and low cost of construction. Total projected area is 5 times the bottom area.

PROBLEM 10 Design a plain sedimentation tank to serve one lakh population @150 litres/head/day, taking a factor of safety of 1.5, detention time of 4 h (approximately), and overflow rate of 20,000 litres/day/m^2 (approximately):

Solution

Volume of the tank = Flow × Detention time

$$= \frac{1.5 \times 100{,}000 \times 150}{1000} \times \frac{4}{24} = 3750 \text{ m}^3$$

Assume a depth of 3 m, then surface area $= \dfrac{3750}{3} = 1250 \text{ m}^2$

As per the surface loading consideration.

$$\text{Surface area of the tank} = \frac{1.5 \times 100{,}000 \times 150}{20{,}000}$$

Hence, $= 1125 \text{ m}^2$

Provide 4 nos. of 36 m × 8 m × 3 m settling tanks. Check for weir loading.

PROBLEM 11 Calculate the diameter (d) of particles that are likely to settle in a plain sedimentation tank (40 m × 10 m × 4 m) with an average flow rate of 1000 m^3/day.

(Sp. gr. of particle = 2.60 and $\mu = 1.00 \times 10^{-2}$ cm^2/s).

Solution

$$\text{Settling velocity } V = \frac{1000}{40 \times 10} \times \frac{1}{24 \times 60 \times 60} \times 100 \approx 0.003 \text{ cm/s}$$

As per Stoke's law,

$$d = \sqrt{\frac{18\,\mu V}{g(s-1)}}$$

or

$$d = \sqrt{\frac{18 \times 1 \times 10^{-2} \times 0.003}{981 \times (2.6 - 1)}} \approx 0.6 \text{ mm}$$

PROBLEM 12 What will be the percentage removal of particles, if the settling velocity of the particle is 0.2 cm/s given by Stoke's law and the critical velocity of flow is 0.3 cm/s given by the overflow rate consideration.

Solution

$$\text{Percentage removal} = \frac{0.2}{0.3} \times 100 = 66\frac{2}{3}.$$

3.3.4 Coagulation

Very small particles of 1–100 μm size (inorganic or organic), causing turbidity do not settle due to gravitational forces only. These colloids are minute, charged particles, have large surface area (more adsorption), exhibit Brownian (erratic, zig-zag) movement and are held in suspension. They have two layers, a primary layer of positive ions and a secondary layer of ions with opposite charge. Colloids are stable due to the zeta potential between these two layers. If this charge or potential is removed and smaller particles are made to combine into a bigger mass (by the addition of a chemical coagulant like alum or ferric salt), settling will take place quickly, easily and completely in a sedimentation tank. Process of coagulation occurs in three stages.

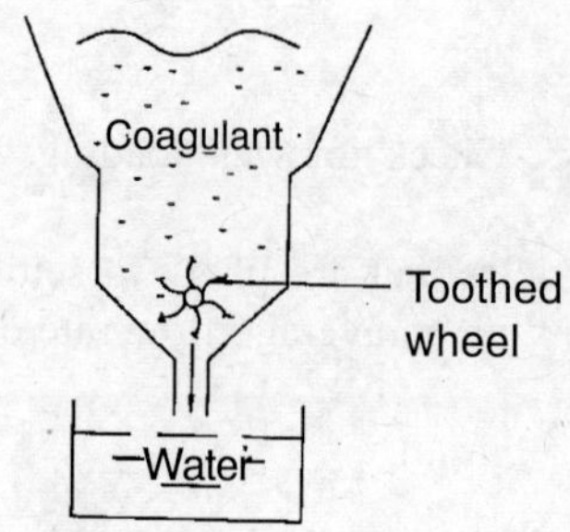

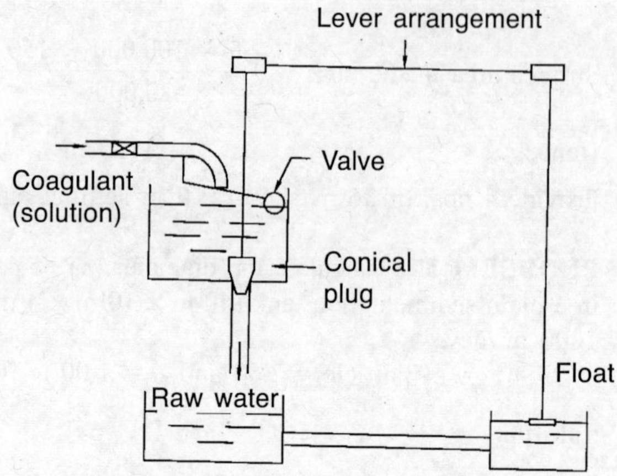

Fig. 3.33 (a) Dry feeding device. **Fig. 3.33** (b) Solution feeding device.

Destabilising

In alkaline waters, alum forms aluminium hydroxide (precipitate), otherwise, $Ca(OH)_2$ or Na_2CO_3 is to be added to impart alkalinity. These Al $(OH)_3$ flocs, i.e. positively charged trivalent Al ions, neutralise the negatively charged colloidal particles. This requires feeding or addition of coagulant to water as shown in Figs. 3.33(a) and 3.33(b). Purpose of feeding is dosing the chemical proportional to the water flow i.e. the more the flow rate, the more is the chemical dose. Rapid (instant) mixing (Fig. 3.34(a)) is the first part of coagulation where good opportunity for contact is given. Formation of hydroxide is needed for coagulation.

Coagulation of turbid water consists of a series of operations that result in the destabilisation of the colloids leading to its removal from the suspending fluid. Ideally it is possible to distinguish four sequential steps in coagulation, viz. (i) dispersion of the coagulant through the liquid, (ii) small floc formation due to nucleation, (iii) flocculation until flocs are large enough for efficient separation from water, and (iv) phase separation by settling.

Aggregation and binding

After the negative charges of turbid particles are neutralised (destablised) by the

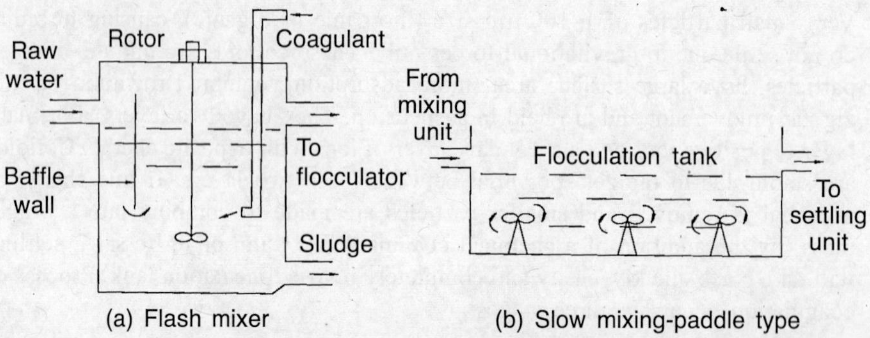

(a) Flash mixer (b) Slow mixing-paddle type

Fig. 3.34 Mixing devices (*Contd.*)

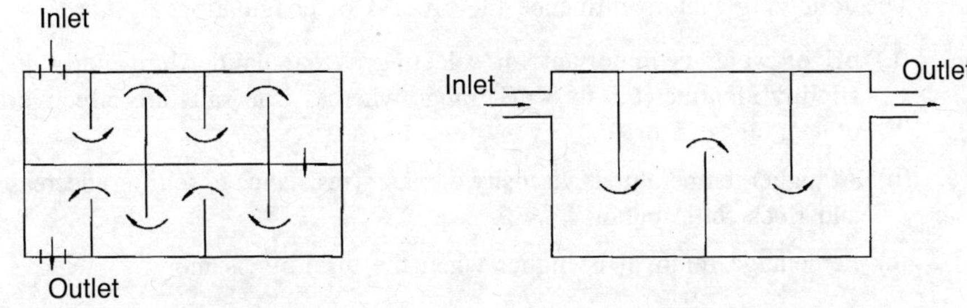

(c) Plan of around the end baffle type mixing basin

(d) Elevation of over and under the baffle type mixing basin

Fig. 3.34 (c and d) Baffle wall chambers (mixing devices).

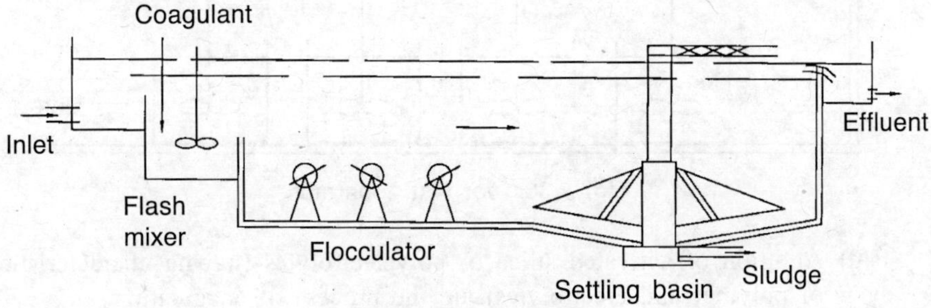

Fig. 3.35 Clariflocculator.

positively charged $Al(OH)_3$, attraction causes aggregation of colloids. These minute flocs entangle some more impurities during floc formation stage, movement and descent, and settle down. Before finally settling, they collect maximum amount of turbid particles from water. So, flocculation is a slow mixing second stage of coagulation process, by which flocs form and combine with impurities into big masses. This is achieved by slow stirring devices (Fig. 3.34(b)). It is to agglomerate destabilised particles under gentle agitation. Liquid movement and particle movement gives opportunity for collision and contact, attach and bind, grow and settle. More collisions may break flocs again. No settling is preferred in a flocculating basin. Mixing basins with baffle walls Fig. 3.34(c) and (d), also serve the purpose of slow mixing.

Flocculation is considered to consist of two stages, namely perikinetic and orthokinetic (from peri, meaning near; ortho, meaning proper, and kinetic, meaning motion). Peri-kinetic flocculation is caused by Brownian or random motion of the particles within water. Orthokinetic flocculation is caused by systematic movements or transport of the particles within the fluid.

Settling

It is a final phase of coagulation process where finer suspended particles formed earlier into a big mass with greater affinity are removed in a settling tank or a clarifier (Fig. 3.35).

The following factors influence the process of coagulation.

(a) pH of water is important for selecting a coagulant. Alum works in a slightly alkaline (6.0 to 8 pH) range whereas iron salts are effective in other ranges (5 or 9).

(b) At higher temperatures, viscosity of water (resistance to settling) decreases and flocs settle better.

(c) Percentage removal is higher when the turbidity is more.

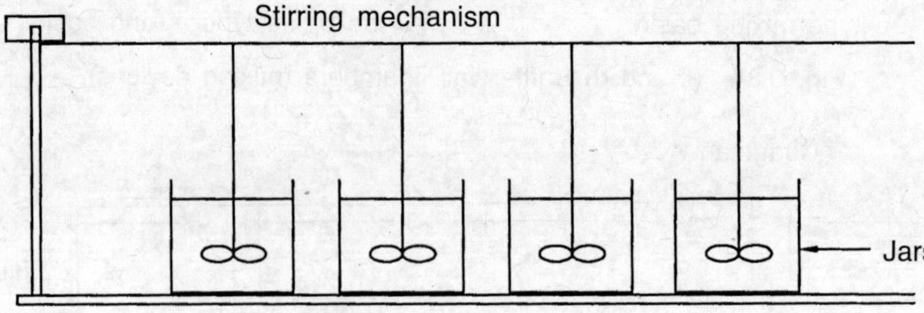

Fig. 3.36 Jar test apparatus.

(d) Addition of activated silica or polyelectrolytes (having characteristics of polymer and electrolytes) aid the process of coagulation.

(e) The dose which gives maximum turbidity removal (optimum dosage) is determined by a Jar test (Fig. 3.36). Different dosages of chemicals are added to water samples in 4 or 6 beakers in a jar test wherein rapid mixing (about 60–80 r.p.m. for 5 minutes) and slow mixing (1–5 r.p.m. for 20–30 minutes) are completed. Flocs are allowed to settle for 20–30 minutes. The dosage which gives the maximum removal is selected. Jar test is useful because it determines (i) which coagulant is useful, (ii) how much mixing time and speed should be allowed, (iii) effects of pH, temperature, etc. on the process of coagulation.

The following chemical reactions take place in coagulation.

$$Al_2(SO_4)_3.18H_2O + 3Na_2CO_3 \rightarrow 2Al(OH)_3 + 3Na_2SO_4 + 3CO_2 + 18H_2O$$

$$Al_2(SO_4)_3.18H_2O + 3Ca(OH)_2 \rightarrow 2Al(OH)_3 + 3CaSO_4 + 18H_2O$$

(As coagulation needs alkalinity of water, additional alkalinity may be added).

$$FeSO_4.7H_2O + Ca(OH)_2 \rightarrow CaSO_4 + Fe(OH)_2 + 7H_2O$$

$$4Fe(OH)_2 + O_2 + 2H_2O \rightarrow 4Fe(OH)_3$$

$$2FeCl_3 + 3Ca(OH)_2 \rightarrow 3CaCl_2 + 2Fe(OH)_3$$

Alum and iron salts (ferrous sulphate, ferric chloride or a combination) are used as coagulants. Iron salts produce heavy, dense, quick settling flocs in a wider range of pH. But alum is preferred as it is highly efficient in the normal pH range of water and is quite economical. In fact, it is better if a floc continues (whirling and confusing) to collect more impurities in water instead of quickly settling.

Design factors

The various parameters for designing the flash mixer, flocculator and clarifier having different detention times are listed below.

(i) Flash mixer — detention time < 1 minute
Flow velocity ~ 1 m/s
Depth: 1–3 m
Mixing speed: 100 r.p.m.

(ii) Flocculator—detention time: 20–30 minutes
Flow velocity: 1 metre/minute
Depth: 3–4 m

(iii) Clarifier—detention time ~ 1½ hours.
Overflow rate: 20–40 $m^3/m^2/day$
Depth: 3–5 m

PROBLEM 13 Find the requirements of alum and lime to treat water (10^7 L/day) at alum dosage (30 mg/L) when original alkalinity present is 8.5 mg/L.

Solution

(i) Alum required $= \dfrac{10 \times 10^6 \ (L/d) \times 30 \ (mg/L)}{1000 \times 1000} = 300 \text{ kg/day}$

(ii) 4.5 mg/L alkalinity ($CaCO_3$) is required for 10 mg/L dosage of alum, i.e. 13.5 mg/L is required. So (13.5 – 8.5), i.e. 5.0 mg/L is to be supplemented.

Since 56 mg of CaO (quicklime) is required for obtaining 100 mg of $CaCO_3$,

$$\text{lime required} = 5 \times \frac{100}{56} \times \frac{10 \times 10^6}{10^6} = 90 \text{ kg/day}$$

3.3.5 Filtration

Water with suspended impurities and bacteria pass through a bed of clean sand and the impurities are retained in the filter layers.

Mechanism

Pores between sand particles act like a sieve and retain bigger particles. Pores act as a miniature settling tanks. Due to physical and electrochemical forces, surface of the sand particles in the filter bed attract and attach (adsorb) turbid particles.

Bacteria utilise (decompose) organic material deposited in the top layers of sand bed and forms a fine (slimy) layer. This layer is very effective in removing finer solids also. More deposition in the top layers attract more bacteria. So bacteria and solids are retained at the top layers only. If the rate of filtration is high and coarser sands are used, suspended solids penetrate into deep layers and are retained. Suspended impurities are transported due to hydrodynamic forces. Water flows nearer to the sand particle surfaces. Particles get attached to the sand surfaces and deposit in the pores. Once deposited, they adhere and are removed from the water flow. Straining, inertial impact, interception, adhesion and adsorption are involved in the process of filtration. As filtration is a continuous process, bed clogging also takes place in deeper layers. This clogged bed offers resistance to flow and is termed as loss of head in filtration. Filters are stopped and cleaned when the loss of head reaches a maximum permissible limit or when impurities are coming into the filtrate.

Types of filters

Water filters are classified according to the following criteria.

(a) Rate of flow per unit surface area—slow filter, rapid filter, constant rate, declining rate

(b) Kind and type of filter media employed—sand, coal (or anthracite) or diatomaceous earth

(c) Direction of flow through the bed—down flow, up flow or bi-flow

(d) Pressure or gravity filters.

In gravity flow filters, two flow rates are adopted, one is a low rate of filtration (100–150 L/h/m^2) known as slow sand filter, the other is a high rate of filtration (5000–6000 L/h/m^2) known as rapid sand filter. Pressure filter maintains high rate (more than 8000 L/h/m^2). Diatomaceous filters, dual media (sand and coal) filters and up flow filters are also adopted. As clogging is limited to top layers of filter only, different ways of utilising lower layers have been tried.

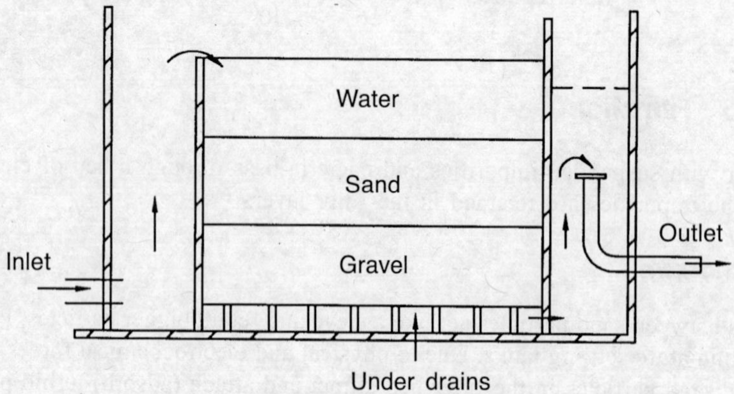

Fig. 3.37 Slow sand filter.

Working of slow sand filter. Figure 3.37 gives details. Large area of land is required for filtration with sand filter and even number of units are preferred in sizes of 60 m × 40 m (approximately). The effective size (i.e. sieve size in mm that permits 10% of sand by mass to pass through) of sand used is 0.25 to 0.35 mm; the uniformity coefficient (i.e. the ratio between the sieve size that will pass 60% and the effective size) is 2 to 3. Due to clogging in top layer in a course of two or three months, filter head loss reaches upto 80–100 cm. Filter is stopped. Top few centimetres of sand is removed for cleaning. New sand is replaced for starting the filter again. The dirty sand is hand cleaned with jets of water, dried and again used later. As the flow rate is less and sand size is small, filter

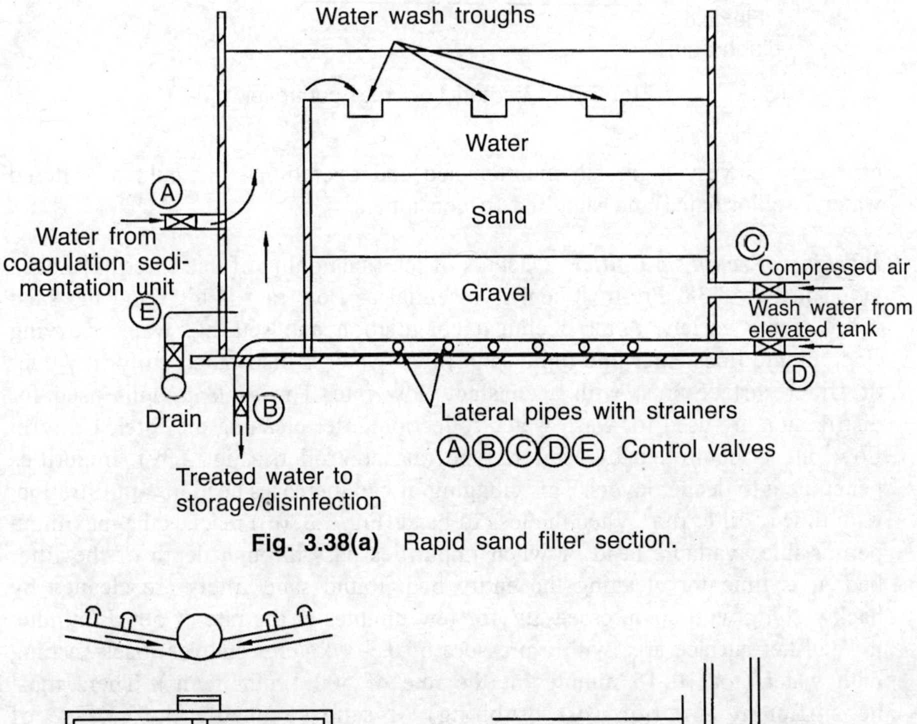

Fig. 3.38(a) Rapid sand filter section.

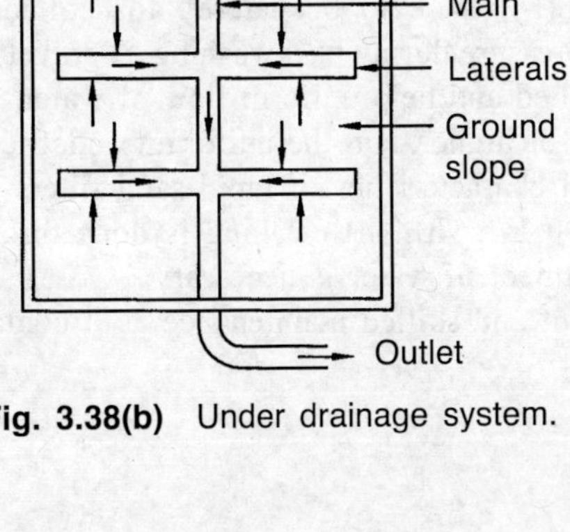

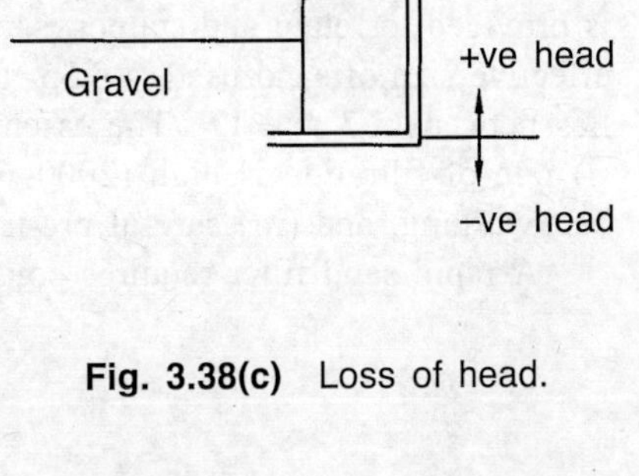

Fig. 3.38(b) Under drainage system.

Fig. 3.38(c) Loss of head.

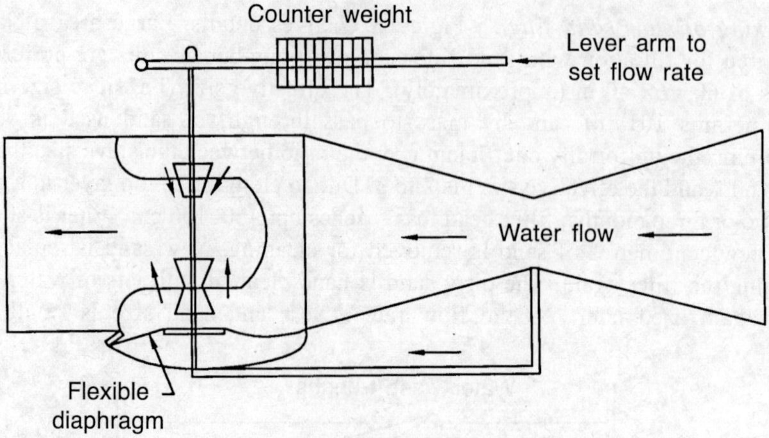

Fig. 3.39 Venturi flow rate controller.

efficiency is very high. No maintenance and operational difficulties. Filtered water is chlorinated and supplied to consumers.

Working of rapid sand filter. Details of the functioning of this filter have been given in Fig. 3.38. Pre-treatment is essential as flow rate is high and the sand is of coarse variety. After chemical coagulation and settling, water carrying fine, minute flocs enter into rapid sand filter. Water is distributed uniformly over the filter surface area, with a constant flow rate. The same troughs used for distribution are used for wash water collection, after cleaning the filter. Uniform flow rate is controlled by Venturi type rate controllers (Fig. 3.39). Impurities penetrate into deeper layers. This clogging of bed increases head loss in filtration, with time of filter run. When the loss of head (Fig. 3.38(c)) reaches the maximum permissible available head or when impurities pass through depth of the filter bed, it is time for cleaning the entire bed. Rapid sand filters are cleaned by backwashing with compressed air (for few minutes at the rate of 500 L/minute/m^2 of filter surface area with a pressure of 0.3–4 kg/cm^2 and (or) backwashing with water (for 10–15 minutes at the rate of 500 L/minute/m^2). There must be sufficient agitation (or scrubbing) of sand to enable the release of impurities clogged. Wash-water gutters are situated at a height over the filter bed (40–50 cm) such that all the impurities are collected towards a drain avoiding loss of sand. Under-drains (Fig. 3.38(b)) have to be catered for collecting the filtrate and distributing water with pressure during backwashing. A gravel bed is provided to retain and support sand bed and help uniform flow of water. The effective size of sand used is 0.35 mm or higher, and the uniformity coefficient lies between 1.3 and 1.7. The essential characteristics of rapid sand filters are: (i) rate of filtration is high (2000–6000 L/m^2/h), (ii) cleaning is done through backwashing, and (iii) careful pre-treatment of water is necessary.

A rapid sand filter requires constant and skilled maintenance as it contains

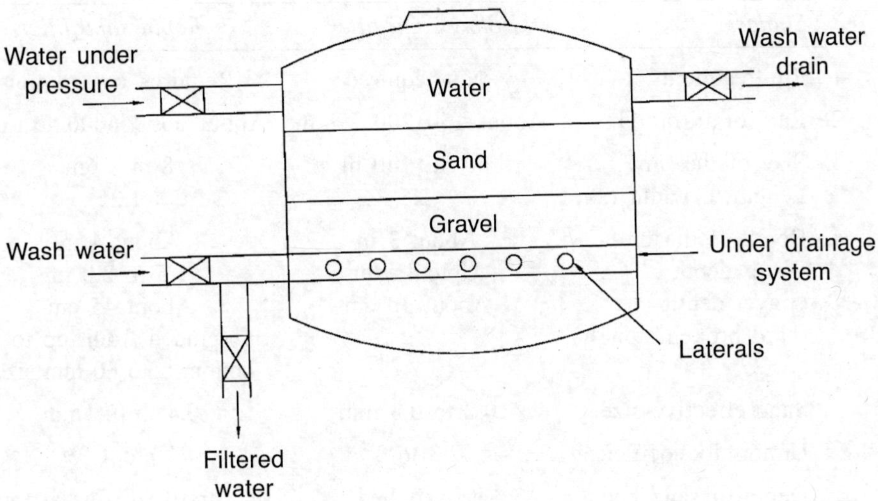

Fig. 3.40 Pressure filter.

a number of control devices such as flow rate controller, loss-of-head indicators, backwashing system and various valves. Rapid sand filters may create serious troubles also. Filter bed may develop cracks or mud balls or incrustations over sand surface or air binding in the pores. All these things reduce the efficiency of filtration. Negative head formation (when loss of head exceeds the permissible value) is very undesirable. Hence cleaning of filter is a highly skilled operation.

Pressure filters (Fig. 3.40) are cylindrical tanks (enclosed) 5 to 6 m in height and 2–3 m in diameter. Filtration rate is very high (8000 L/h/m^2) in this type of filters due to maintenance of 3–4 kg/cm^2 of pressure. These are just similar to rapid sand filters but are more compact units used in industrial sectors.

PROBLEM 14 Determine the size of a rapid sand filter unit for filtering 0.15 ML/hour of water.

Solution

Assuming a rate of filtration of 6000 L/h/m^2,

$$\text{area of the filter required} = \frac{0.15 \times 10^6}{6000} = 25 \text{ m}^2$$

Assuming a length to width ratio of 1.25, we get

Length = 6 m and width = 4.5 m

Comparison of Filters

Subject	*Slow sand filter*	*Rapid sand filter*

(*Contd.*)

Subject	*Slow sand filter*	*Rapid sand filter*
1. Pre treatment	Not required	Requires coagulation
2. Rate of filtration	About 150–200 L/h/m^2	About 5000–6000 L/h/m^2
3. Size of the unit	60 m × 40 m	8 m × 6m
Length to width ratio	1.5	1.25
4. Depth of the unit	About 3 m	About 4 m
Water depth	Upto 1.5 m	1.5 to 2.0 m
Gravel depth (hard broken stone)	About 30 cm	About 45 cm (graded from top to bottom upto 60 mm size)
Sand effective size	0.2 to 0.3 mm	0.4 to 0.6 mm
Uniformity coefficient	2.3 to 3	Upto 1.75
Depth of sand bed	Mixed bed 70 to 80 cm	Stratified, fine at top and coarse at bottom (about a metre in depth).
5. Under-drainage system	Split lateral tile drains (not involved in cleaning filter)	Perforated drains, laterals (10–15 cm diameter) and main (15–30 cm) with holes about 8–10 mm size at 125 mm centre to centre (useful for backwashing)
6. Loss of head (not greater than)	1 m	2 to 3 m
7. Cleaning interval	Once in two or three months.	Once in a day or two days
8. Method of cleaning	Top 5 cm sand removed, washed dried and replaced	Backwashing with compressed air and water for 10–15 minutes, allowing sand to rise for 50–60 cm.
Wash water requirement	Small	2–3% of water filtered.
9. Efficiency	Colour, turbidity, bacteria removal is better in slow sand filter but water from rapid sand filter is also good.	

3.3.6 Disinfection

The main aim of disinfection is to destroy pathogenic (disease causing) bacteria and thus prevent water borne diseases. Chlorine is used universally as a disinfectant, normally after water filtration. Addition of chlorine to water results (hydrolysis) in the formation of HOCl and HCl, and later HOCl ionises into H & OCl (free available chlorine). Chlorine combines with ammonia (chloramines), resulting in combined available chlorine. As pH of solution increases, more HOCl dissociates into OCl ion. When pH value is less than 5, only elemental chlorine exists. Above pH of 10, mostly OCl remains. Hypochlorite is more powerful in the disinfection process. Chlorine penetrates into the cell wall and inactivates the bacterial cell metabolism. Rate of kill, a first order reaction (efficiency of disinfection), depends upon:

(a) dosage (amount of chemical used and its concentration)

(b) nature and number of organisms

(c) time of contact

(d) water temperature, alkalinity and pH

(e) presence of other compounds.

Chlorine is applied in the solid state (addition of $CaOCl_2$, i.e. bleaching powder) into small water tanks. Chlorine gas in solution (i.e. in the liquid form) is passed through the municipal water mains. It is always preferable to keep a residual chlorine amount of 0.2 to 0.6 mg/L to prevent further contamination in the distribution system. Chlorine demand is the amount of chlorine required with contact time of 30 min., to kill all microscopic organisms, to oxidize or react with chemical compounds and includes the residual amount desired (0.2 mg/L). Heavily conta-

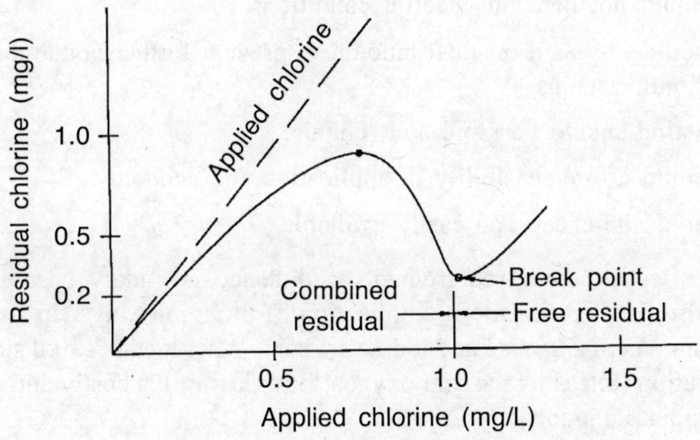

Fig. 3.41 Break point chlorination.

minated waters need super dose of (excess) chlorination, i.e. about 5–10 mg/L. Dechlorination is to be followed later by the addition of SO_2 or $Na_2S_2O_3$.

The dosage of chlorine is decided by available chlorine in the disinfectant and chlorine demand. In a laboratory, different dosages of chlorine are added to a number of water samples taken from the same source. A graph (Fig. 3.41) is drawn between the applied chlorine dose and residual chlorine in each sample. Applied chlorine is used for the formation of chloro compounds initially and later for killing bacteria when dose reaches a certain amount. After these two purposes are satisfied, the residual chlorine is proportional to the applied chlorine amount, once again as shown in the graph. Break point chlorination is defined as the amount of chlorine at which nearly all the residue is free chlorine (H and OCl). Application of chlorine at or slightly higher than the break point concentration will remove unpleasant tastes and odours, will have adequate bactericidal effect and will leave a desired residual chlorine. The dose of chlorination necessary to reach break point will vary widely. Normally dosage is selected in such a way to maintain the residual chlorine of about 0.2 mg/L.

PROBLEM 15 Calculate the amount of bleaching powder required to treat 10^7 L/day of water with chlorine dosage of 0.8 mg/L.

Solution

Assume that available chlorine in bleaching powder is 40%.

$$\text{Amount of bleaching powder required} = \frac{10^7 \times 0.8}{1000 \times 1000 \times 0.4} = 20 \text{ kg/day}$$

A good disinfectant should fulfil the following requirements:

(1) It should not be organism specific.

(2) It should not render the water toxic.

(3) It should not demand specific conditions.

(4) It should stay as a residual amount to prevent further contamination in distribution mains.

(5) Its action should be rapid and reliable.

(6) It should afford feasibility in application and economy.

(7) It should be cheap and easily available.

Chlorine satisfies all these requirements and hence is widely used in water purification. Boiling (sterilisation), application of ultraviolet rays (in swimming pools), potassium permanganate and Iodine are some other methods of disinfection. Ozone application (releasing nascent oxygen for oxidation) is costly and does not leave any residual amounts.

3.3.7 Other Methods

In addition to the conventional units of sedimentation, filtration and chlorination, depending upon the need, the following units are also incorporated.

Aeration

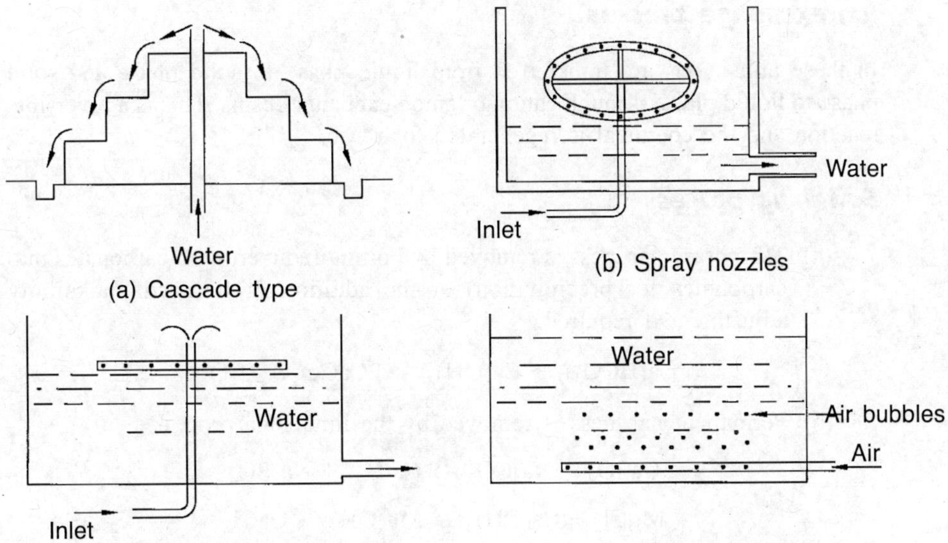

(a) Cascade type (b) Spray nozzles

(c) Surface aerator (d) Diffused air

Fig. 3.42 Aeration systems.

Aeration of water removes bad tastes and odours and freshens water with oxygen. It removes iron by oxidation and precipitation.

Iron may be present in water in one or more forms such as ferrous bicarbonate, ferrous sulphate or organic chelated iron. Manganese is also a constituent in ground waters as manganese bicarbonate.

Hard water responds well for Fe and Mn removal by aeration. But when the water is soft or has a low alkalinity or the iron and manganese are combined with organic matter, lime and soda ash are used to raise the alkalinity to the point necessary for oxidation reaction and coagulants are used to expedite settling. Chemicals like Cl_2 or $KMnO_4$ may be used to release Fe from organic matter.

$$4Fe + O_2 + 10H_2O \rightarrow 4Fe(OH)_3 + 8H$$

Undesirable gases H_2S, CO_2, NH_3, CH_4, etc. are expelled from water by the addition of oxygen.

In cascade type of aeration, (Fig. 3.42) water flows as thin sheets on horizontal steps. About 0.02–0.03 $m^2/m^3/h$ is the area normally provided. In spray nozzles (Fig. 3.42(b)), water droplets sprinkle into the air absorbing oxygen. About 2–3 cm diameter of nozzles may make up for a flow of 20–30 L/h discharge. In diffused aeration system, compressed air is sent into water tanks through diffuser pipes. About 0.5 to 0.8 liter of air per litre of water is diffused in a tank with about 15-45 minutes detention time.

In all these simple devices, surface area of exposure is increased for more oxygen absorption. The rate of absorption also depends upon the temperature, and the difference between the existing concentration and the equilibrium concentration of oxygen in solution.

In the water treatment plant, aeration unit presents a pleasant aesthetic appearance.

Ion exchange process

In these units, ions are transferred from liquid phase to solid phase and solid phase to liquid phase through cation or anion exchange resins. This is a reversible reaction and the chemical is regenerated for reuse.

Softening process

(a) Temporary hardness is removed by boiling (converting bicarbonates into carbonates for precipitation) or the addition of lime with alkalinity adjustment if required.

$$Ca(HCO_3)_2 + Ca(OH)_2 \rightarrow 2CaCO_3 + H_2O$$

(b) Permanent hardness is removed by the lime soda process.

$$CaSO_4 + Na_2CO_3 \rightarrow CaCO_3 + Na_2SO_4$$

$$MgCl_2 + Ca(OH)_2 \rightarrow Mg(OH)_2 + CaCl_2$$

$$CaCl_2 + Na_2CO_3 \rightarrow CaCO_3 + 2NaCl$$

(c) Zeolites (Na_2Z or Na-Ca-Al-Silicate) are used for the removal of hardness efficiently and completely.

$$Na_2Z + Ca \rightarrow CaZ + 2Na$$

By passing brine solution, the Zeolite is regenerated. Synthetic exchange resins (Na_2R) are used in softening process. Ion exchange methods are used for the removal of cations and anions.

PROBLEM 16 Determine the lime and soda requirements to remove hardness from water. (Data-water quantity 1 millon litres, Carbonate hardness 150 mg/L as $CaCO_3$, non-carbonate hardness 90 mg/L as $CaCO_3$)

Solution

Lime requirement can be worked out as follows:

100 mg/L of $CaCO_3$ requires 56 mg/L of CaO

240 mg/L of total hardness needs $\frac{240 \times 56}{100} = 135$ mg/L of lime,

i.e., 135 kg for treating 1 ML of water.

Soda requirement:

100 mg/L of $CaCO_3$ requires 106 mg/L of Na_2CO_3

90 mg/L of non-carbonate hardness needs $\frac{106 \times 90}{100} = 95.4$ mg/L of soda i.e. 95.4 kg for treating 1 ML of water.

The following data will be useful while calculating the above requirements when hardness is present due to $MgSO_4$ and $CaSO_4$.

(a) 120 mg/L of $MgSO_4$ requires 56 mg/L of CaO

(b) 120 mg/L of $MgSO_4$ needs 106 mg/L of Na_2CO_3

(c) 44 mg/L of CO_2 needs 56 mg/L of CaO

(d) 136 mg/L of $CaSO_4$ requires 106 mg/L of Na_2CO_3

Adsorption

A number of water pollutants diffuse on to the surface of a solid adsorbent. Organic chemicals (imparting colour, tastes/odours) heavy metals, fluorides, etc. are removed using solid adsorbents like activated carbon, activated alumina and silica gel.

Large surface area accumulates more ions. Adsorption capacity depends upon solvent-solute-solid system. The affinity and attraction are due to (i) electrostatic forces (ion exchange as a result of charge), (ii) Physical forces (due to VanderWaal's), (iii) Chemical interactions (surface attachments). The rate and amount of adsorption are controlled by (i) diffusion of sorbate from bulk solution on to the surface of sorbant, (ii) sorbate gets attached on the surface, (iii) sorbates gets accumulated in the micropores.

The amount of sorbate adsorbed per unit mass of solid adsorbent is a function of concentration of solute remaining in solution at equilibrium conditions. After equilibrium no more transfer takes place. Isotherm specifies the equilibrium conditions between how much can be adsorbed and how much is remaining in solution. These isotherms are expressed by different theories in Freundlich, Langmuir, and B.E.T. equations. Adsorption is a simple phenomena and a specific unit operation for removing some liquid or gaseous water pollutants.

Activated carbon is inexpensive, used in a wide range, and large quantities of water can be treated efficiently.

Membrane process

Dissolved salts are removed (in specific cases only) in reverse osmosis and electrodialysis units (separated by a membrane between two different solutions).

Reverse osmosis

Liquid flows from a concentrated solution to dilute solution across a semi-permeable membrane of polymeric material (due to application of an external pressure or a driving force) which allows only water to pass through and not salts from solution.

Electrodialysis

When water flows through an array of membranes (cationic and anionic), various ions migrate towards electrodes, leaving behind pure water.

Corrosion control

Corrosion means deposition of substances on the pipe material, and chemically reacting with the material, resulting in the loss and reduction of life of pipe. Acid waters are more corrosive. Corrosion is controlled by (i) removal of CO_2 from water, (ii) adjustment of pH, (iii) application of protective coatings on pipes, (iv) insulation and cathodic protection.

Fluorides are removed by (i) the addition of alum and lime followed by clarification (ii) passing through beds of activated alumina, (iii) ion exchange process through special resins

Salinity is removed by a distillation (evaporation) and condensation of water vapour, electrodialysis, reverse osmosis, ion exchange processes.

Tastes and odours are removed by adsorption with activated carbon. *Nitrates* are removed by biological decomposition. *Iron salts* are removed by oxidation and precipitation. *TDS* are removed by evaporation/distillation process.

3.3.8 Rural Water Supply

Protected water supply system in a village consists of a water storage tank having few water taps for collection by individual consumers. For population of 1000 persons (at the rate of 40–50 L/head/day), 5-m diameter tank with 3-m height is sufficient. Bleaching powder (1 kg/day) is mixed once a day.

SUMMARY

- Water is available for drinking purposes as surface water (rivers and lakes) and ground water,
- Various chemical and bacteriological impurities are found in water—more suspended impurities in surface waters and more dissolved salts in ground water.
- Drinking water is purified from safety considerations as per the permissible standards.
- Per capita demand of water is taken as 150–250 litres/head/day, considering domestic, public, commercial and other uses in a community.
- Water is collected by intakes (surface water) tube wells (ground water) and is distributed to the consumer after adequate treatment.
- The entire system of pipes, valves, pumps and all appurtenances involves huge expenditure (more than treatment costs) and hence requires proper design.
- Suspended solids are removed by sedimentation, coagulation and filtration processes.
- The process of coagulation is due to addition of coagulants such as alum

whereas sedimentation allows the particles to settle by natural gravitational force.

- Slow sand filters are quite useful for removing suspended solids and harmful bacteria. High rate gravity sand filters are operated in urban areas successfully to purify water.
- Chlorination is a must for killing harmful microorganisms before supplying water to the consumer.

REVIEW QUESTIONS

1. Compare the surface water and ground water for their quality, quantity and collection aspects.
2. List out the various impurities associated with water quality resulting in water pollution.
3. Explain the term *per capita demand.* How do you arrive at an average rate of water demand for the purpose of design in water supply and purification.
4. Write a note on the advantages and disadvantages of various water distribution methods and layouts.
5. Explain the principles involved in the process of coagulation.
6. Design plain settling tanks for treating 2 million litre per day (MLD) of water.
7. Distinguish between a slow sand filter and a rapid sand gravity filter from construction and operational considerations.
8. Why is chlorination preferred for disinfection of water supplies?
9. Write a note on the various methods available for removing permanent hardness.
10. Write short notes on:
    (a) Drinking water quality standards
    (b) Water pollution
    (c) Intake wells
    (d) Hardy cross method
    (e) Jar test
    (f) Theory and mechanism of filtration
    (g) Break point chlorination
    (h) Aeration

OBJECTIVE TYPE QUESTIONS

A. ***Each of the following questions comprises four alternatives. Select the most appropriate answer for each.***

1. Which of the following is not a water borne disease?
   (a) Polio (b) Jaundice
   (c) Malaria (d) Typhoid
2. Amount of water required for extinguishing fire is stored in:
   (a) Fire hydrants (b) Water mains
   (c) Service reservoirs (d) intake wells
3. Maximum permissible concentration of nitrates in drinking water is (in mg/L)
   (a) 250 (b) 60
   (c) 45 (d) 30
4. Which of these valves is fixed on the delivery side of a pump set?
   (a) Gate valve (b) Air relief valve
   (c) Reflux valve (d) Venturi meter
5. Turbidity is caused due to
   (a) Settleable matter (b) Dissolved impurities
   (c) Colloidal matter (d) Passage of light
6. Cement concrete pipes are joined by
   (a) Expansion joint (b) Flanged joint
   (c) Spigot & socket joint (d) None of the above.
7. Alum dosage of 40 mg/L is equivalent to
   (a) 4 kg/m^3 (b) 0.4 kg/m^3
   (c) 0.04 kg/m^3 (d) 2 kg/m^3
8. pH value of N/100 Nitric Acid is
   (a) 1 (b) 3
   (c) 2 (d) None of the above.
9. Intake wells are provided to
   (a) Carry water from a treatment plant
   (b) Draw water from a ground source
   (c) Collect water from a surface source
   (d) Distribute water to the consumer
10. In Hardy Cross Method, corrections are applied to
    (a) Pipe diameter (b) Loss of head
    (c) Flow quantity (d) Directions of flow
11. Flocculators are
    (a) Sedimentation units (b) Instantaneous mixing devices
    (c) Slow stirring mechanisms (d) Chemical feeding systems

12. Flow rate controller is associated with which unit of water treatment
    (a) Sedimentation (b) Coagulation
    (c) Filtration (d) Disinfection

13. A chemical used to control algal growths in reservoirs is
    (a) Alum (b) Bleaching powder
    (c) Copper sulphate (d) Activated carbon

14. Surface area of a rapid sand filter to treat 104 m^3/day of water is approximately
    (a) 100 m^2 (b) 60 m^2
    (c) 80 m^2 (d) 120 m^2

15. High value of factor of safety is considered while designing pipes of a particular type of diameter (in distribution of water)
    (a) big (b) medium
    (c) small (d) any size

16. In which layout system of distribution, more number of valves and fittings are required?
    (a) dead end (b) gravity
    (c) grid iron (d) all the above

17. After break point chlorination, chlorine remains as
    (a) free residual (b) combined residual
    (c) free and combined residual (d) none

B. ***Fill in the blanks with suitable words:***

1. Infiltration galleries are located near ________.
2. Desirable pressure in distribution mains should be ________.
3. The possibility of sewage contamination is indicated in drinking water, by the presence of ________.
4. In comparison with surface water, ground water is usually rich in ________ and poor in ________.
5. Gas formation in the fermentation tubes in B.Coli test indicates ________.
6. In calculating the capacity of a service reservoir, mass curve is drawn between time and ________.
7. Lime is added in the coagulation process to ________.
8. 50 mL. of water sample requires 5 mL. of N/20 Na_2CO_3, to completely precipitate Ca as $CaCO_3$. Water hardness (in terms of $CaCO_3$ equivalent) is ________.
9. More nutrients in lake waters cause ________.
10. pH affects the process of coagulation and ________ in water treatment.
11. Infiltration galleries are packed with gravel to prevent the entrance of ________.
12. Efficiency of settling in a sedimentation tank, doesn't mainly depend on ________.
13. Rapid sand filter needs incoming nature of water to be

CHAPTER 4

Domestic Sewage

4.1 SEWERAGE SYSTEM

4.1.1 Introduction

Water, used for domestic purposes, collects impurities and becomes waste water. Thus domestic waste water (sewage) consists of faecal matter and used water from baths, kitchen, etc. (sullage). This waste water (99% of water and 1% solids) is transported through a network of sewers, away from residential units. Sewage contains human excreta which is the main cause for water borne diseases. Decomposed organic matter emanating foul gases and odours, and many other impurities contaminate water sources and land surfaces. Water stagnation leads to unhygienic conditions. Collection, conveyance, treatment and disposal of sewage (and storm water) from a domestic environment is of primary importance. Municipal or government organisations must take care of this. Various process industries also discharge their untreated liquid effluents into the land and water bodies. This poses major environmental challenges. Industries, therefore, are expected to take responsibility for treating their effluents before the final disposal.

4.1.2 Methods of Sanitation

In a conservancy system, human excreta is collected by a human agency, which is a very unhealthy practice. Its storage and transportation create more nuisance. Sullage water, if unutilised for the kitchen garden, flows in open drains or percolates into the ground. Refuse or dry solid waste is collected through garbage bins, transported by lorries and dumped in low-lying areas away from residential areas. Thus, in a conservancy system, human excreta, sullage and rubbish are collected, conveyed, treated (sometimes), and disposed separately. They are not mixed with one another. This conservancy method is unhygienic and is responsible for spreading communicable diseases.

In a water carriage system, adequate (15–20 litres) quantity of water is added to human wastes in a water closet, and the sewage flows (99% water) in closed conduits (sewers) without creating any unsightly and unhygienic conditions in the locality. Sullage waters and shredded garbage are allowed to join the sewage flow. Liquid wastes are treated to reduce their foul odour, and ultimately disposed safely into water or land bodies. This water carriage system effectively eliminates most of the water borne diseases in the community. In building planning, toilets need not be located away from the dwelling units. The water carriage system has many advantages but requires huge financial investment. However, cost should not be a constraint since human welfare in a society is of prime concern.

Separate and combined systems

Sewerage system is a network of pipes (sewers) carrying liquid wastes, consisting of a number of laterals (small diameter pipes located in the streets) and mains (connecting a number of laterals) as shown in Fig. 4.1. Trunk sewer carries flow from a number of mains and through the outfall sewer, sewage is disposed into natural water courses. The network pattern depends upon the land topography and the layout of colonies and streets. The entire sewage should be collected and diverted to one place, i.e. to a treatment plant. The diameter of sewer increases as the sewage is collected from more and more residential units. The diameter of trunk sewer is more than that of mains. The sewage treatment plant is located at *a lower level site* (as sewage flows by gravity), outside the city residential units, and closer to the disposal point (land or surface water source). Once a city or a town is having a sewerage system, storm water may be allowed to flow into the sewers in combination with sewage; or storm water may be separately collected in open drains or closed pipes and disposed off more easily and safely.

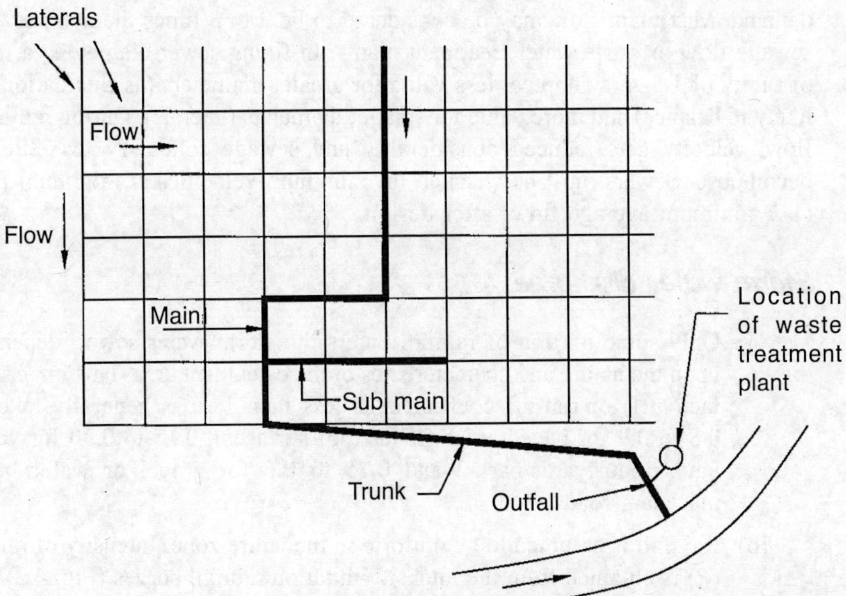

Fig. 4.1 Sewerage system.

A combined system is preferable when

(a) Dilution of sewage is advantageous

(b) The rainfall is uniform throughout the year and

(c) It is not feasible and economical to have two separate sewers.

However, combined sewerage system requires pipes of larger diameter. It is difficult to maintain minimum (self cleaning) velocities at times when there is no storm water flowing into the sewers. Heavy rains may block the entry of house drainage into the sewers. Normally a separate sewage system excluding the rain water is preferred. Depending upon the convenience and economy, rain water may be allowed into domestic sewers at suitable locations.

4.1.3 Quantities of Sewage and Storm Water Flow

Per capita sewage contribution

The quantity of sewage (without storm water), is known as dry weather flow in a sewer and is almost equal to the quantity of water (100–250 litres/head/day) supplied to the populations (including private water sources). Losses in the water supply system are compensated by ground water infiltration into the sewerage system. The rate of infiltration depends upon ground water table and soil characteristics. Safely, 80%–90% of average rate of demand of water supply can be assumed to be per capita sewage contribution. The fluctuations and the factors affecting the sewage flow are just similar to those affecting per capita water demand. Maximum flow may be considered to be 2 to 3 times the average daily sewage flow in waste water treatment plants. In fixing sewer diameters, a factor of safety of 1.5–4 is adopted, less value for greater diameters (as fluctuations are likely to balance) and more value for smaller diameters (laterals). During minimum flow, velocity gets reduced considerably and sewage solids try to settle and decompose. Sewers must be checked for minimum velocities at minimum flows (1/2 minimum average flow) after design.

Storm water discharge

(a) Only some portion of rainfall enters the storm water sewer, depending upon the nature and characteristics of the catchment area. This impervious factor (I) or run-off coefficient is less than 1. It is generally taken as 0.8 to 0.9 for paved areas (roads/roof surfaces), 0.15 to 0.30 for vacant lands (courtyards/parks) and 0.25 to 0.6 for gravel or water bound macadam roads.

(b) Rainfall is assumed to be uniform in the entire zone. Intensity of rainfall (R) is obtained from the intensity-duration rainfall curves (Fig. 4.2) over a certain period of years. These curves can also be represented in the form of an equation:

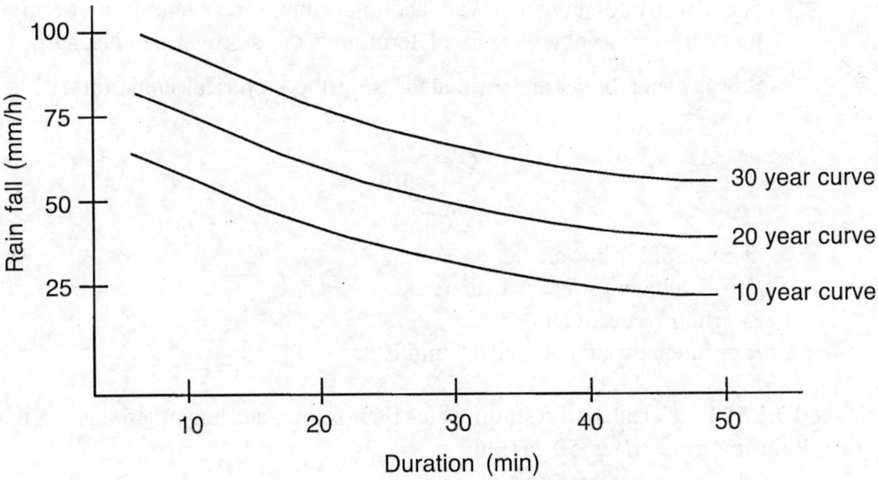

Fig. 4.2 Intensity-duration rainfall curves.

$$R = \frac{a}{T + b}$$

where

R = rain intensity in mm/h

T = time of duration of rainfall in minutes

a and b are constants approximately taken as

a = 760 and b = 10 for 5–10 min. duration of storm, and

a = 1020 and b = 10 for 20–100 min. duration of storm.

(c) At the beginning of a storm, water flows from the adjoining areas into sewers (Fig. 4.3) *AB*, *BC* and *CD*. If rain continues for some more time, flow from sewer *AB* (zone 1) reaches sewer *BC*, in addition to the flow (from zone 2) into sewer *BC*. Similarly, flow from *AB* and *BC* moves into sewer *CD*, in addition to flow from zone 3 into the sewer *CD*. There is also a possibility that storm water from zone 1 from *AB* may reach *BC*, after water from *BC* moves into *CD* or when *CD* receives water from *BC*, after water leaves the sewer *CD*. It only means that the diameter of storm sewer has to be decided considering the time of entry from a particular zone into that storm water sewer and the time of flow from

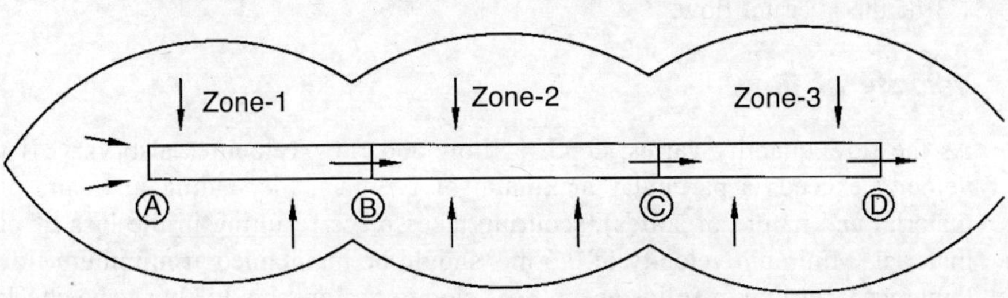

Fig. 4.3 Time of concentration.

one sewer into another sewer. The total time of concentration at a given point has to be obtained for determining the storm water discharge.

(d) Storm water flows are estimated using the empirical equation

$$Q = \frac{AIR}{360}$$

where

Q = runoff in m^3/s.
A = catchment area in hectares
I = runoff coefficient
R = Intensity of rain fall in mm/h.

PROBLEM 1 Calculate the storm water flow from a catchment area, given that:
Rain intensity (R) = 50 mm/h
Area (A) = 54 hectares and
30% of area consists of roof with runoff rate as 0.9,
30% of area consists of open field with runoff rate as 0.2,
40% of area consists of roads with runoff rate as 0.4.

Solution The net impervious coefficient I, i.e.

$$\text{Runoff rate} = \frac{0.30 \times 0.9 + 0.30 \times 0.2 + 0.40 \times 0.4}{0.3 + 0.3 + 0.4} = 0.49$$

$$\text{Storm water flow } Q = \frac{AIR}{360} = \frac{54 \times 0.49 \times 50}{360} = 3.675 \text{ m}^3\text{/s}$$

4.1.4 Sewer Design

The following parameters are considered in sewer design.

Quantity of flow

If 200 L/head/day is the amount of water supplied to a community with one lakh population, quantity of sewage is taken as 3 times the average rate as the maximum flow, i.e. 3 × 200 × 100 m^3/day. Also the minimum discharge is taken as 1/2 or 1/3 of the average flow.

Velocity of flow

As the flow quantity varies, depth of flow and flow velocities also vary. If the velocity exceeds a particular maximum of 2–3 m/s, (depending upon the pipe material and nature of solids), scouring takes place resulting in the loss of pipe material. Minimum velocity of 0.6 m/s should be maintained at minimum flows. Otherwise, settling of solids occur. So, velocity of flow should always be checked for minimum and maximum flows, as self cleaning velocity has to be maintained without fail. If a flow velocity of 0.8 m/s is taken during design for full depth, it is likely to be within safe limits as sewage always flows through a sewer keeping it partially full.

Diameter

Sewers always run partially full (1/2 to 3/4th depth) so as to allow ventilation for the foul gases liberated during the decomposition of sewage. For different depths of flow (for different flow quantities), the flow velocities vary. These variations between d/D, q/Q and v/V are related as shown in Table 4.1, based upon the depth and the velocity relations in the Manning's or Chezy's equations. These are useful in checking the flow velocities under all circumstances of flow quantity and depth.

Table 4.1 Hydraulic Elements of Circular Sewers

Sl. No.	d/D	q/Q	v/V
1	0.20	0.088	0.615
2	0.40	0.337	0.902
3	0.50	0.500	1.000
4	0.60	0.671	1.072
5	0.70	0.726	1.120
6	0.80	0.978	1.140

Notes:

1. When the sewer is running half full (i.e. $d/D = 0.5$) its hydraulic mean depth $\left(\frac{1}{2}\frac{\pi d^2}{4} \div \frac{\pi d}{2} = \frac{d}{4}\right)$ and flow velocity $\left(\frac{Q}{2} \div \frac{1}{2}\frac{\pi d^2}{4}\right)$ are equal to those values when the sewer is running full
2. If Q is 0.08 m^3/s while running full, then $q = 0.978 \times 0.08$ for $d/D = 0.8$
3. These values may also be represented in the form of curves

Slope

Sewers are laid (1–4 m) below ground level with slope of 1 in 100 to 1 in 1000 for the wastes to flow under gravity. The sewer gradient depends upon the natural slope of ground, the flow velocity and the discharge. Deeper excavations are avoided and sewage is lifted up and allowed to flow again with a required slope. Flushing tanks are used for cleaning the sewers when ground slope is not adequate for maintaining the self cleaning velocity. Sewers at different levels can be joined through drop manholes.

Slope is the hydraulic gradient h/L, calculated as the loss of head per 1000 m.

length of pipe. For a known quantity of the flow and required velocity of the flow, suitable diameter and slope needed are calculated using a nomogram (Fig. 4.4) based on Manning's formula (for sewers flowing full). Join Q and V by a straight line to obtain d and s values.

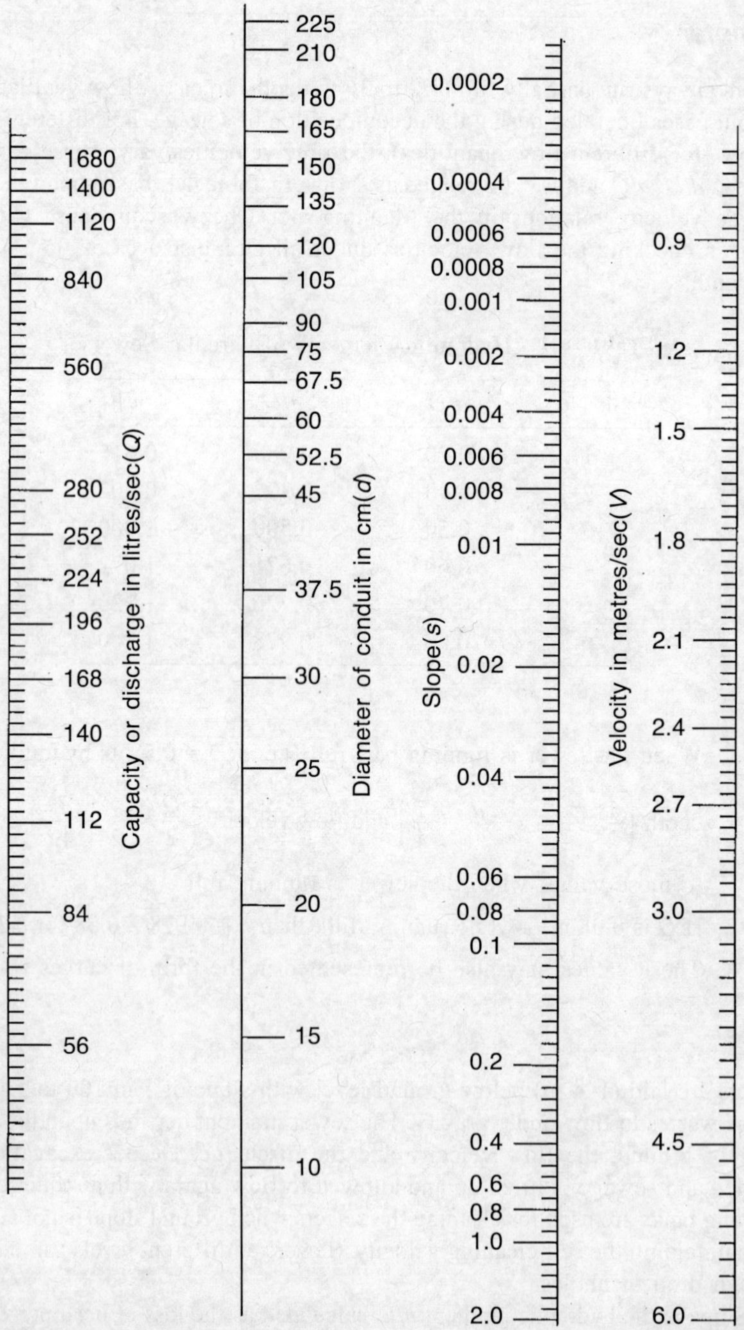

Fig. 4.4 Nomogram based on Manning's formula for $n = 0.013$ for sewers running full.

All these parameters are related in different equations, as stated below.

1. $V = \frac{1}{n} r^{2/3} s^{1/2}$ (Manning's formula)

2. $V = cr^{1/2}s^{1/2}$ (Chezy's formula)

where

V = velocity of flow (m/s)

r = hydraulic mean radius = $\frac{\text{Area of cross section}}{\text{Wetted perimeter}} = \frac{\pi d^2}{4} \div \pi d = \frac{d}{4}$

(d = diameter (m), for circular sewer, flowing full).

s = hydraulic gradient (h/L) = $\frac{\text{difference in level}}{\text{length of the sewer}}$

c = Chezy's constant (50–80, depending upon size, shape, roughness, etc. of the pipe)

n = rugosity coefficient (0.012 to 0.014, depending upon the nature of the material of the pipe).

3. $\frac{h}{L} = \frac{fV^2}{2\,gd}$ Darcy-Weisbach formula.

4. $Q = AV = V \cdot \frac{\pi d^2}{4}$

where Q = sewage flow (m^3/s)
h = head loss (m)
f = friction coefficient
g = gravity constant, 9.81 m/s^2

PROBLEM 2(a) Determine the velocity of flow and discharge (flowing full) in a sewer, given that, diameter of sewer is 60 cm and Slope, 1/500 or 0.002.

Find the change in velocity if the flow is half full.

Solution

(i) Using Chezy's formula, we have $V = 60\sqrt{\left(\frac{0.60}{4}\right)(0.002)} = 1.04$ m/s

and $$Q = \pi\,\frac{(0.6)^2}{4} \times 1.04 = 0.28 \text{ m}^3/\text{s}$$

(ii) Using Manning's formula, $V = \frac{1}{0.013}\left(\frac{0.60}{4}\right)^{2/3}(0.002)^{1/2} = 0.86$ m/s

$$Q = \pi\,\frac{(0.60)^2}{4} \times 0.86 = 0.24 \text{ m}^3/\text{s}$$

(iii) Using Hazen William's formula,

$$V = 85\left(\frac{0.60}{4}\right)^{0.63} (0.002)^{0.54} = 0.89 \text{ m/s}$$

$$Q = \pi\,\frac{(0.60)^2}{4} \times 0.89 = 0.25 \text{ m}^3\text{/s}$$

(iv) If $d/D = 0.5$; $v/V = 1$ and $q/Q = 0.5$; hence the result.

PROBLEM 2(b) Determine the slope and diameter of a sewer, if

$$Q = 1500 \text{ L/s and } V = 1.5 \text{ m/s.}$$

Solution

$$\text{Cross sectional area} = \frac{\pi d^2}{4} = \frac{Q}{V} = \frac{1.5 \text{ m}^3\text{/s}}{1.5 \text{ m/s}} = 1 \text{ m}^2$$

$$d = \sqrt{\frac{4}{\pi} \times 1} = 1.12 \text{ m}$$

Using Chezy's formula,

$$1.5 = 50\sqrt{\frac{1.12}{4} \cdot s}$$

Slope = 0.0032

PROBLEM 2(c) Calculate the diameter and discharge of a circular sewer laid at a slope of 1 in 400 when it is running half full and with a velocity of 1.9 m/s

Solution

Using Manning's formula (p. 99), we have

$$1.9 = \frac{1}{0.012}\left(\frac{d}{4}\right)^{2/3} \times \frac{1}{\sqrt{400}}$$

or $$d = 1.23 \text{ m}$$

$$\text{Discharge} = \pi \times \frac{1.23^2}{2 \times 4} \times 1.9 = 1.13 \text{ m}^3\text{/s}$$

PROBLEM 2(d) Determine the diameter and the velocity of flow, if

$$Q = 0.5 \text{ m}^3\text{/s, and } s = 1/500$$

Solution

According to Darcy-Weishbach head loss formula,

$$\text{Slope} = h/L = \frac{fV^2}{2\,gd} = \frac{1}{500} = 0.002$$

i.e.,

$$\frac{V^2}{d} = \frac{2g \times 0.002}{0.01} = 3.91 \qquad \text{(i)}$$

Further, $$Q = AV = \frac{\pi d^2}{4} V, \quad \text{or,} \quad Vd^2 = \frac{4Q}{\pi} = \frac{4 \times 0.5}{\pi} = 0.64 \qquad \text{(ii)}$$

Solving (i) and (ii) we have

$$V = 1.73 \text{ m/s and } d = 0.76 \text{ cm}$$

4.1.5 Sewer Construction

The physical structure required for the collection and disposal of both sanitary sewage and storm water sewage should be constructed according to certain parameters such as the material used for the pipes, the shape of the sewers and the method of laying the sewers.

Materials

Sewer materials must withstand corrosion and abrasion, they should have sufficient strength and durability. Stoneware pipes (< 40 cm dia.) are available in smaller lengths and used for smaller diameter sewers (laterals). They are easy to handle, to lay and can be easily jointed. They are cheap and durable. They have smooth surface and are corrosion resistant. Asbestos cement pipes (50 to 100 cm dia.) are suitable for domestic sanitary fittings.

Concrete pipes are strong and economical for higher diameters but are likely to be affected by chemical action. Cast iron pipes (150 to 750 cm dia.) are costly and preferred for outfall sewers or used under special circumstances.

Shapes

Circular sewers are economical and are suitable for small and medium diameters (both for laterals and mains). Egg shaped sewers are preferred for combined system (sewage + storm water). They maintain self cleaning velocities even when the flow is less. Rectangular or horse shoe type is more stable. They can be cast at the site when used for larger diameters. Open drains are used for the disposal of sullage water (from domestic use) or rain water (Fig. 4.5).

Laying of sewers

The following steps are involved:

1. It is essential to maintain proper slopes, as the sewage flows under gravity only.

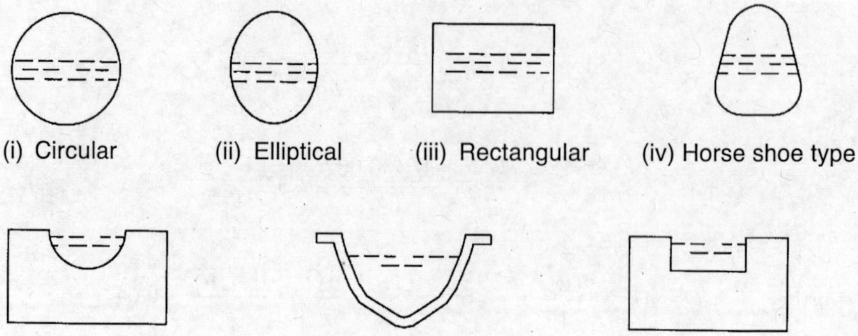

(v) Open drains

Fig. 4.5 Shape of sewers.

2. Sewers are laid on the foundation beds to avoid settling.
3. If necessary, the ground water should be pumped out and the sides of the trenches should be provided with timber planks and struts.
4. The invert level of the sewer has to be properly verified.
5. Sewers are laid, jointed and tested for leakage before the trench is filled back with excavated earth.
6. To avoid deeper excavations, sewage may be lifted (pumped) to higher elevations and again allowed to flow by gravity.
7. Flushing tanks are provided to maintain the self cleaning velocities (if slope is not sufficient) depending upon the topographical features.
8. Manholes are provided at regular intervals, bends, junctions, etc., so that inspection, cleaning and the general maintenance can be done with ease.
9. Care must be taken to eliminate the possibility of cross connections to the water mains.
10. Sewers are likely to contain toxic gases like H_2S, CO_2, CH_4, SO_2, NH_3, etc. which may cause asphyxiation or explosions. Sufficient care must be taken to provide ventilation during inspection.

4.1.6 Sewer Appurtenances

These are the structures constructed at suitable locations along a sewerage system.

1. *Manholes* are chambers giving access to sewers for inspection and cleaning. They are provided at every bend, junction, change of gradient or diameter. On straight reaches also manholes are spaced at every 50–100 m, depending upon the sewer diameter. Drop manholes are provided when two sewers meet at different levels. They are shown in Fig. 4.6.
2. A *street inlet* (catch pit) is located along the road side at intervals of 50 m or so, to collect rain water from roads, into sewers (Fig. 4.7).

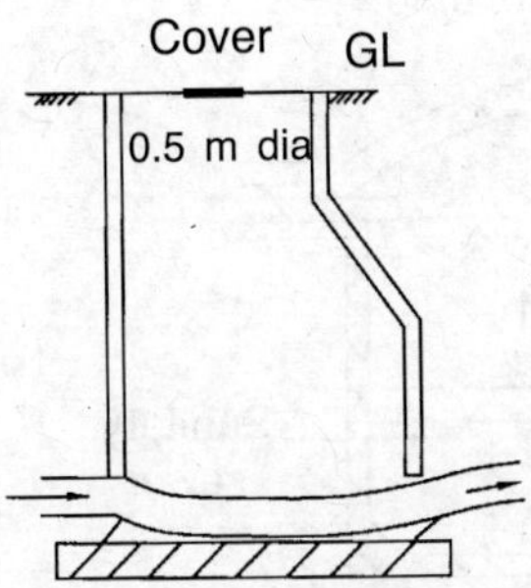

Fig. 4.6(i) Manhole

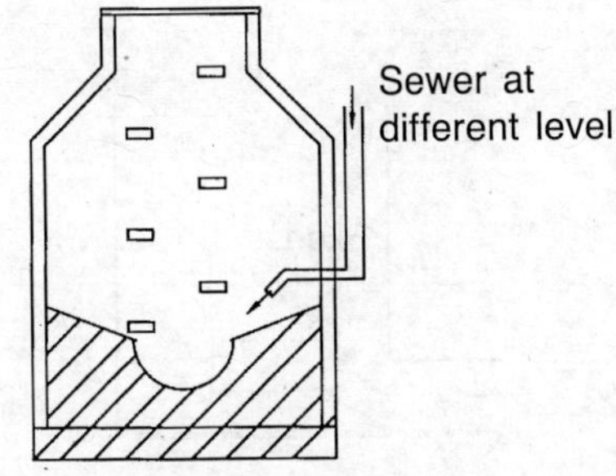

Fig. 4.6(ii) Drop manhole

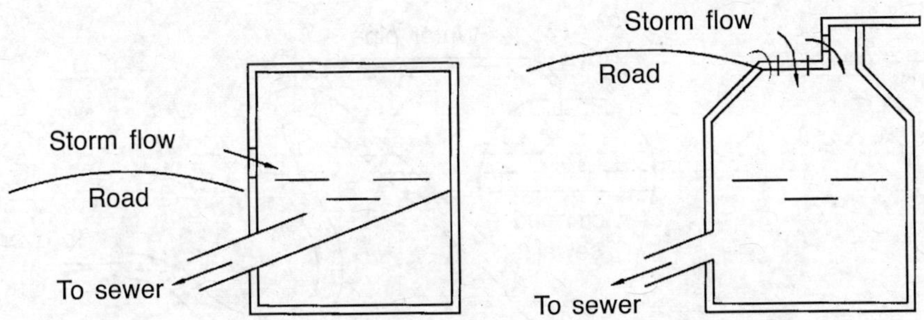

Fig. 4.7 Street inlets.

3. *Grease and oil traps* (Fig. 4.8) are provided to remove grease and oil from sewage from hotels, petrol bunks, etc., before admitting into the sewer line.

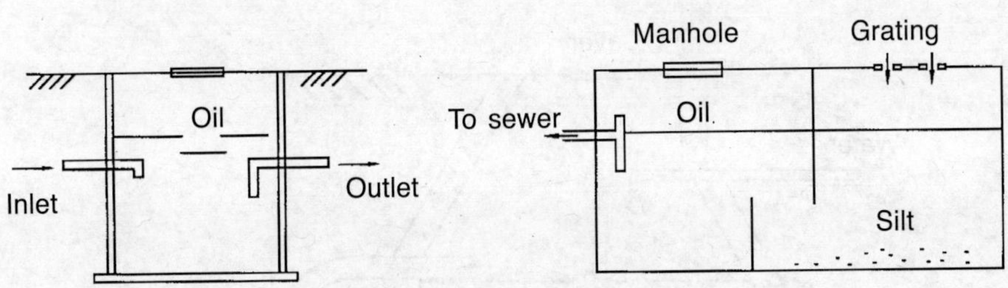

Fig. 4.8 Grease and oil traps.

4. *Storm water regulators* (Fig. 4.9) are provided to divert the excess storm water from the sewer.
5. *Flushing tanks* (Fig. 4.10) are used to clean the sewer periodically. When water reaches a specified level in the tank, Bell is lifted up under air pressure from below. Then water moves into the sewer line, because of siphonic action.
6. *Inverted siphons* (Fig. 4.11) are provided when a sewer has to carry flow below the hydraulic gradient during ground depression for a passing road or a railway line. It runs full under gravity such that no silting takes

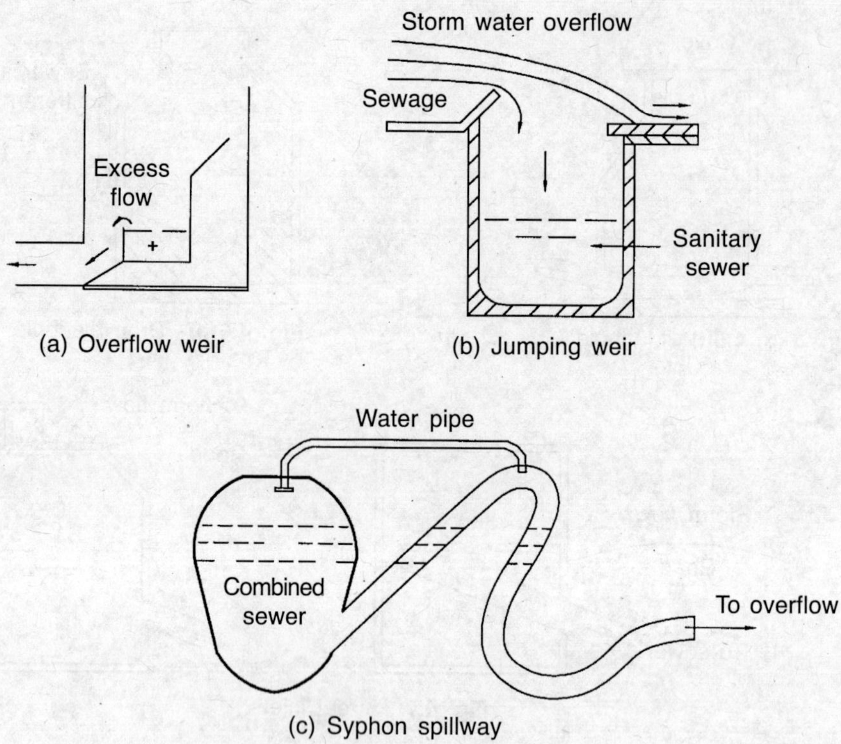

Fig. 4.9 Storm water regulators.

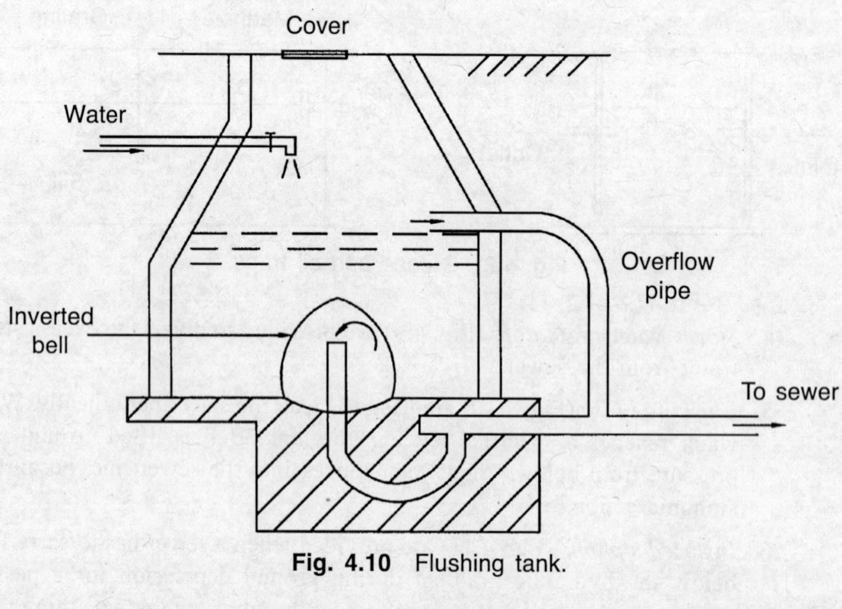

Fig. 4.10 Flushing tank.

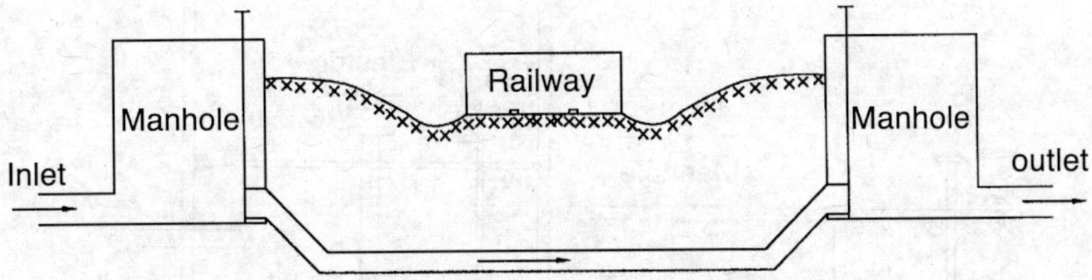

Fig. 4.11 Inverted siphon.

place. The velocity of flow should be atleast 0.8 m/s, even during minimum flow. Two or three pipes of different diameters are provided for taking different types of discharges.

7. The sewage is automatically lifted up to about 5 m by *Shone's ejector* (Fig. 4.12). When the sewage level reaches a certain level, air entrapped in the upper cup, lifts the spindle and the lower cup thus giving more opportunity for the sewage to move out of the ejector with adequate pressure. Once the ejector chamber is emptied, sewage quantity retained in the lower cup pulls down the spindle back to position.

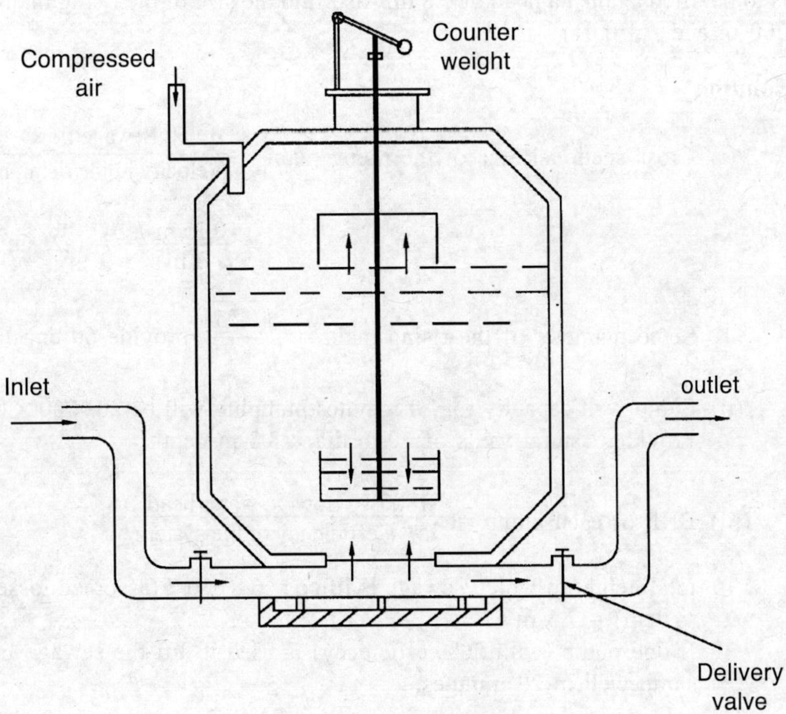

Fig. 4.12 Shones ejector.

8. *Centrifugal pumps* are used to lift the sewage (containing solids and harmful chemicals). A typical pumping station is shown in Fig. 4.13.

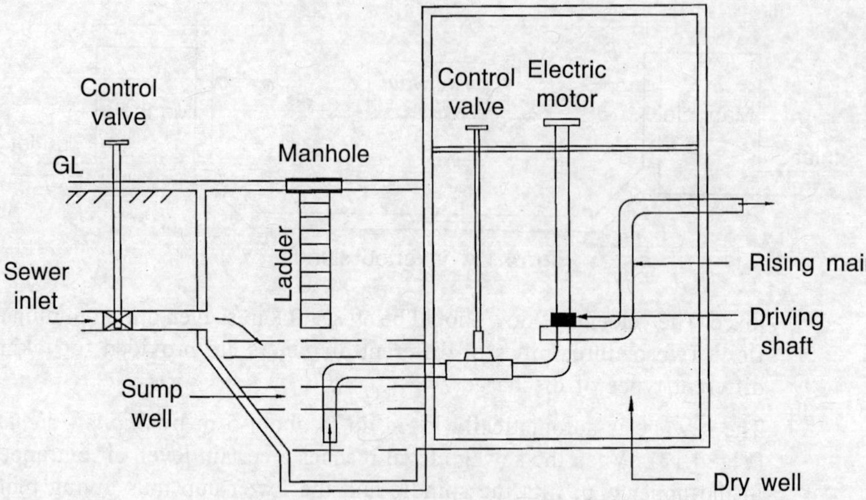

Fig. 4.13 Pumping station.

PROBLEM 3 Determine the BHP of a pumping unit if sewage flow of 0.25 m^3/s is to be lifted against a head of 6.8 m. Also find the size of the rising main (length 200 m and $f = 0.01$)

Solution

(i) Cross sectional area of the rising main $= \dfrac{\text{Sewage flow}}{\text{Velocity in the rising main}}$

$$= \frac{0.25 \text{ m}^3/\text{s}}{1 \text{ m/s}} = 0.25 \text{ m}^2$$

Since diameter of the rising main $= \sqrt{\dfrac{0.25}{\pi/4}}$, provide 60 cm diameter.

(ii) Sump well capacity for 20 minutes pumping will be $20 \times 60 \times 0.25$ m^3. Provide 2 sump wells of 10-m dia × 3.3-m depth

(iii) BHP of each pump $= \dfrac{1000 \times \text{flow} \times \text{water head}}{75 \times \text{efficiency of the system}}$

total height to which sewage is lifted = 6.8 + 0.7 (assumed losses in the system) = 7.5 m

If one pump (with 50% efficiency) is used to lift the sewage from one sump well in 20 minutes,

$$\text{BHP of each pump} = \frac{1000 \times 0.20 \text{ m}^3/\text{s} \times 7.5 \text{ m}}{75 \times 0.50} = 40 \text{ HP}$$

4.1.7 House Drainage

Figure 4.14 shows the plan of a building with water supply, waste water and rain water connections. The general requirements of a sanitary plumbing system can be listed as follows.

1. Waste water from latrines, urinals, sinks, bathrooms, wash basins, etc. should always flow by gravity, automatically with self cleaning velocity. The requirements for sanitary installations depend upon the number of persons using them and the type of building in which they are being installed, viz. a cinema hall, hotel, bus stand, hostel, or a factory or any other building.

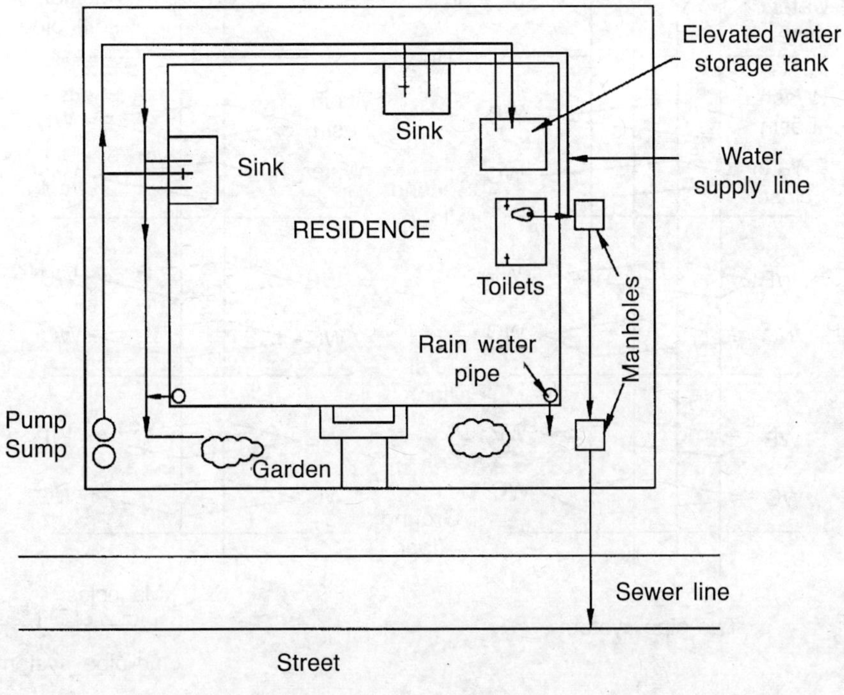

Fig. 4.14 House drainage system.

2. All pipes should be laid preferably on the outer walls, or underground locations. This gives easy access for repair and escaping of foul gases (if any).
3. Manholes should be located at suitable places.
4. The outfall house sewer should be at an elevated level to prevent the back flow from the public sewer.
5. The plumbing system should have traps (water seal) to avoid foul smells. The intercepting trap prevents the entry of foul gases into the house sewer from the public sewer.

6. Adequate ventilation eliminates the foul gases.
7. All drains must be water tight. All fixtures must be of suitable size. Cross connections between the water mains and the soil pipes should be avoided at all costs.
8. Location of various units must be economical. There has to be a provision for easy drainage and compactness.

Figure 4.15 depicts one-pipe and two-pipe systems of plumbing in a multistoried building.

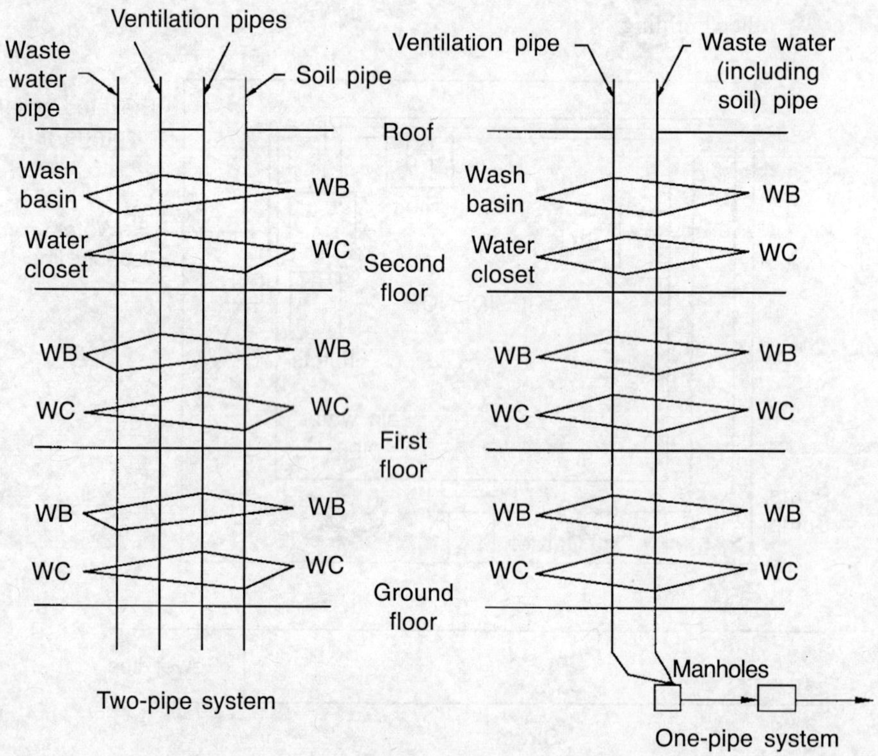

Fig. 4.15 One-pipe and two-pipe systems.

In a two-pipe system, there are two sets of vertical pipes Fig. 4.15 one for draining from water closets and the other for draining waste water from sinks, wash basins, bathrooms, etc. Soil and waste pipes, are provided with separate ventilation shafts.

In a one-pipe system, only one pipe collects human excreta and sullage water. It is also provided with a ventilation pipe. In both the cases, ventilation pipes may be separately provided or the same pipe may be extended upto the top of the roof level for the purpose of ventilation.

Figure 4.16 shows the various important fittings used in a plumbing system.

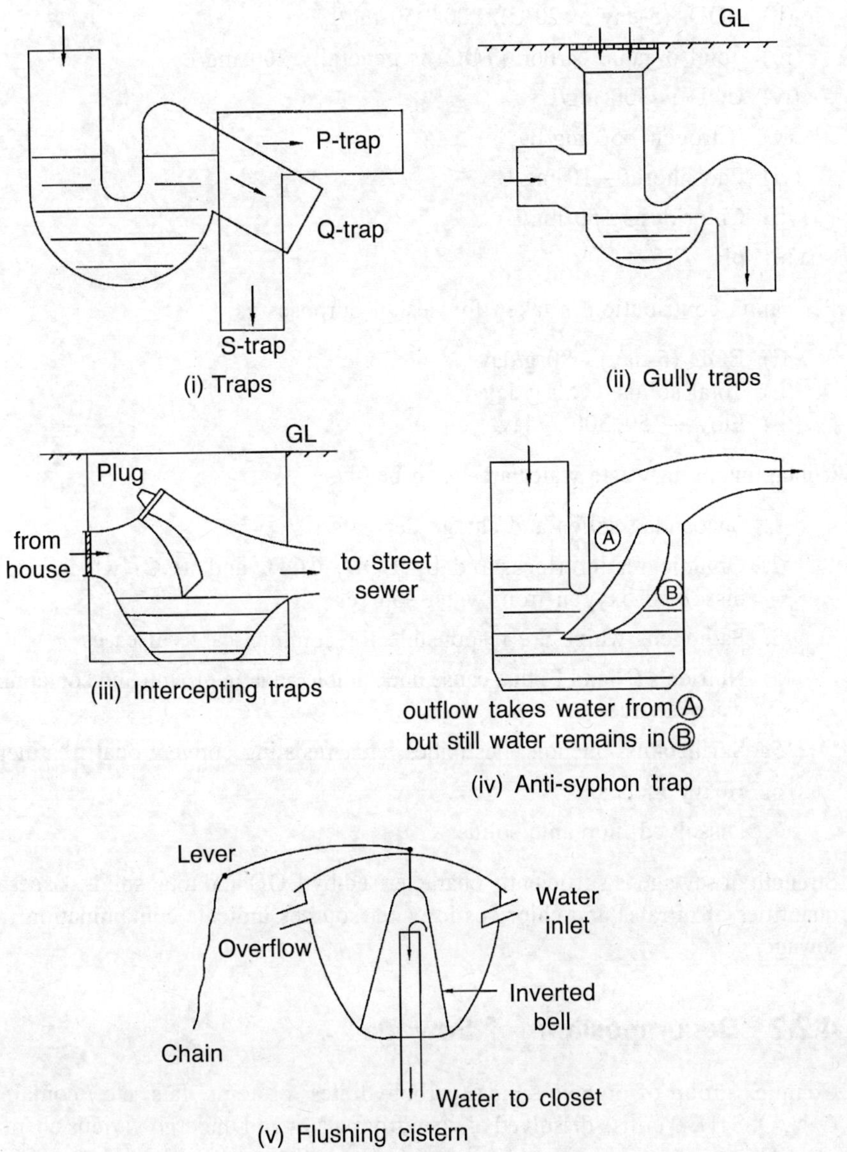

Fig. 4.16 Plumbing fixtures.

4.2 SEWAGE CHARACTERISTICS

4.2.1 Sewage Composition

Sewage contains more than 99% of water and less than 1% solids (by weight). It's average composition is given as follows.

(i) Total solids are about 1000 mg/L including organic, inorganic, settleable and volatile solids (dissolved—600 mg/L and suspended—400 mg/L).

(ii) BOD, (5-day at 20°C), 200–250 mg/L
(iii) Total organic carbon (TOC) is generally 200 mg/L
(iv) COD is 300 mg/L
(v) Nitrogen—50 mg/L
(vi) Phosphorus—10 mg/L
(vii) Chlorides—150 mg/L
(viii) pH—7.5

Per capita contribution is taken for design purposes as:

(i) BOD (5-day)—80 g/day
(ii) Total solids—250 g/day
(iii) Flow—150–300 L/day

Contaminants in waste water are listed below.

1. Suspended solids and sludge deposits
2. Organic matter (measured by BOD, COD, and TOC) which deplete dissolved oxygen from water sources
3. Pathogens which are responsible for communicable diseases
4. Nutrients (N and P) that cause undesirable aquatic growth and contaminate water sources
5. Surfactants, phenols, pesticides, etc., resisting conventional treatment
6. Heavy metals
7. Dissolved inorganic solids

Strength of sewage is sufficiently characterised by BOD and total solids. Excessive quantities of nitrates and chlorides in water sources indicate contamination from sewage.

4.2.2 Decomposition of Sewage

Complex group of organic solids (carbohydrates, proteins, fats, etc.) containing C, N, O, S, H, P utilise dissolved oxygen from water and undergo various chemical reactions due to the action of microorganisms (bacteria, algae, protozoa, etc). This process of conversion of the complex organic substances into more stable compounds is known as decomposition of sewage. Sewage always contains organic matter and bacteria. The organic matter serves as the food for bacteria. So, biochemical decomposition is a continuous process to reduce the strength of sewage. During the decomposition, gases escape and solids settle easily. The liquid becomes less harmful in course of time. The process of decomposition which utilises the dissolved oxygen is known as aerobic decomposition. The aerobic bacteria work in the presence of dissolved oxygen. The final end products of aerobic decomposition are carbon dioxide and water. Carbonates, sulphates, nitrates, etc. are also likely to be present in the liquid effluent.

The decomposition in the absence of dissolved oxygen is known as anaerobic decomposition. The anaerobic bacteria decompose sewage into organic acids, CO_2, H_2S, NH_3, CH_4, mercaptans (organics with SH group), humus, etc. as end-products.

Facultative bacteria can live both in aerobic and anaerobic conditions. Microorganisms are active under two specific temperature ranges such as mesophylic range, i.e. 20–45°C, and thermophylic range, i.e. 45°–60°C.

The rate of biochemical reactions is determined by the amount of dissolved oxygen utilised in aerobic reactions and the amount of gases liberated in anaerobic conditions. When aerobic reactions are over and dissolved oxygen is not available, anaerobic reactions take place. So, this decomposition or conversion of complex organic compounds by microorganisms or transformation into a stable product is a continuous process.

Biochemical decomposition needs the following:

- Organic matter as food for bacteria
- Microbial growth to utilise organic solids
- Dissolved oxygen for aerobic reactions
- A treatment unit giving an opportunity of contact (reactor) and providing favourable conditions of pH, temperature, etc. for the biochemical reactions to take place.

Differences between aerobic and anaerobic reactions are tabulated below:

Aerobic decomposition	*Anaerobic decomposition*
1. Biological oxidation process, nitrification (NH_3 to NO_3)	Hydrolysis of carbohydrates into soluble organics, proteins into amino acids
2. End products are CO_2 and water	End products are CH_4, H_2S, NH_3, etc
3. Aerobic bacteria need oxygen (dissolved in water or from atmosphere)	Anaerobic bacteria work without oxygen
4. After decomposition, liquid effluent may be used for irrigation purposes	Methane has fuel value and sludge is used as manure
5. No offensive odours and foul gases generated	Bad odours emitted
6. Few hours of detention time required	Detention time extends upto 30 days or more
7. More complete reaction	Incomplete reaction
8. Stable	Unstable, requires further treatment
9. Measured from dissolved oxygen utilisation or BOD satisfied.	Measured from gas production or remaining volatile organics

(Contd.)

	Aerobic decomposition	Anaerobic decomposition
10.	Treatment units include trickling filter, activated sludge process, oxidation pond, etc.	Treatment units includes sludge digestion, septic tank, anaerobic lagoon
11.	Energy consuming	Energy yielding
12.	Less sensitive process	Highly sensitive process

4.2.3 Biochemical Oxygen Demand (BOD)

The amount of oxygen required by the organic matter in sewage to undergo the aerobic decomposition (needed for the biochemical reactions) is known as BOD, i.e. biochemical oxygen demand at a given temperature.

BOD reaction takes place in two stages. In the first stage the carbonaceous matter is oxidised in about five days. In the second stage, nitrogenous matter gets oxidised for a prolonged duration. Normally a five-day BOD is determined, from the amount of oxygen consumed in the first five days of biodegradation with proper dilution. Thus, BOD_5 in the sample is equal to the difference between the DO present in the sewage sample immediately after addition of sewage to water, and the DO present in the sample after five days multiplied by the dilution factor. (Note: Samples are diluted in such a way that oxygen remains after the five-day period).

The BOD tests are conducted at a temperature of 20°C for a five-day period to standardize the procedure. The BOD curves are shown in Fig. 4.17.

Mathematical representation. A BOD reaction is a first order chemical reaction involving the utilisation of organic matter by microorganisms. That is, the rate at which organics are utilised is proportional to the amount of organics available, or

$$\frac{dLt}{dt} \; \alpha \; - L_t$$

where L_t = oxygen utilised after time t. Integrating, we get

$$L_t = L_0 e^{-kt}$$

where L_t and L_0 are oxygen (organic matter) equivalents remaining at time t and initial time $t = 0$. k = reaction constant (negative sign indicates decrease). If Y_t represents BOD_t of waste waters stabilised (i.e., $L_0 - L_t$), then

$$Y_t = L_0 - L_0 e^{-kt} = L_0 (1 - e^{-kt}) = L_0 (1 - 10^{-kt})$$

where k is a rate reaction constant which can be determined by Thomas method as illustrated in Problem 5.

For any other temperature, the rate reaction constant is obtained as

$$k_T = k_{20} (1.050)^{T-20}$$

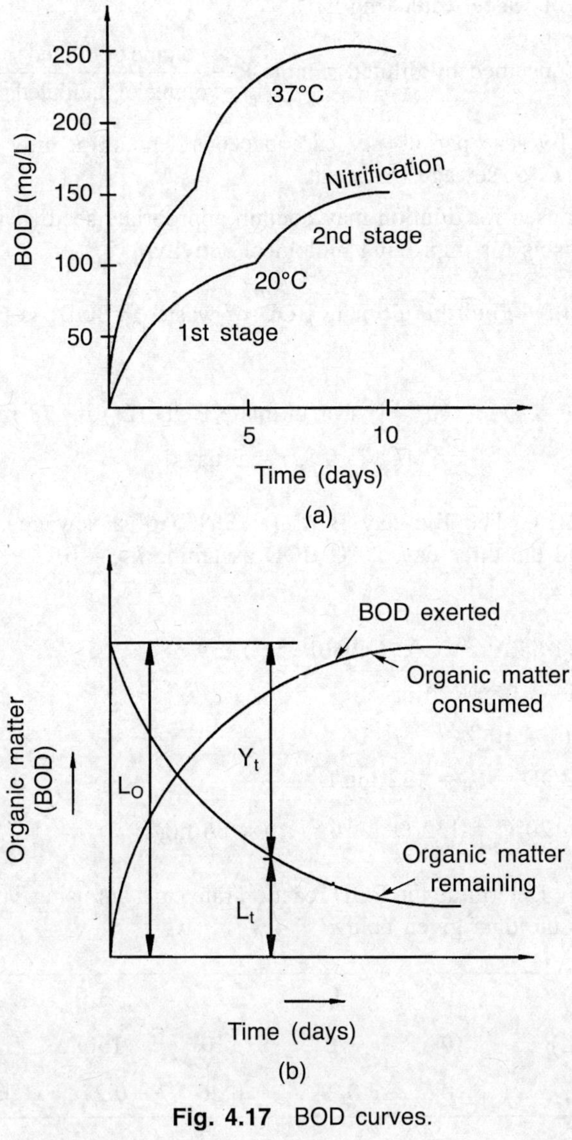

Fig. 4.17 BOD curves.

It may be noted that

1. Ultimate BOD(L_0) is a fixed quantity.
2. The rate of chemical reaction depends upon time and temperature and other favourable conditions.
3. If BOD value at a certain temperature and time period is known, ultimate BOD can be obtained. Then BOD value for any time and temperature is determined using the rate reaction constant for that temperature.

4. BOD of waste water (mg/L) =

$$\text{DO Consumed by diluted sample} \times \frac{\text{volume of diluted sample}}{\text{volume of undiluted sample}}$$

5. BOD five-day period (i.e., carbonaceous demand) is only considered, in all cases of sewage treatment.
6. Water used for dilution may contain appropriate seed (if required) with organisms for improving biological activity.

PROBLEM 4(a) Find the ultimate BOD of waste, if BOD_5 is 600 mg/L

Solution

$$600 = L_0\ (1 - 10^{-0.1 \times 5}) \text{ or ultimate BOD } (L_0) = 878 \text{ mg/L}$$

$$[k_{20} = 0.1,\ t = 5 \text{ days}]$$

PROBLEM 4(b) The five-day BOD (at 30°C) of a sewage sample (Y) is 110 mg/L. Find the three-day, 20°C BOD assuming $k_{20} = 0.1$.

Solution

$k_{30} = k_{20}\ (1.050)^{T-20} = 0.1(1.050)^{10} = 0.159$

$Y_5 = L_0\ (1 - 10^{-k_{30} \times 5})$ or

$110 = L_0\ (1 - 10^{-0.159 \times 5})$

Ultimate BOD = L_0 = 132 mg/L

Now, Y_3 at 20°C = $132\ (1 - 10^{0.1 \times 3}) = 66$ mg/L

PROBLEM 5 Determine the BOD reaction rate constant using Thomas method (Fig. 4.18) for the data given below.

t = time (days)	0	1	2	3	4	5
y = BOD(mg/L)	0	70	110	160	185	200
$(t/y)^{1/3}$	0	0.24	0.26	0.27	0.28	0.29

Solution

Draw graph $(t/y)^{1/3}$ vs. time (t) and obtain intercept and slope as

$$a = 0.23 \text{ and } b = 0.011$$

According to Thomas method,

$$(k) \text{ rate reaction constant} = 2.61\ \frac{b}{a} = 0.125 \text{ per day.}$$

$$\text{Ultimate BOD} = \frac{1}{2.3\,k(a)^3} = \frac{1}{2.3(0.125)\,(0.23)^3} = 290 \text{ mg/L}$$

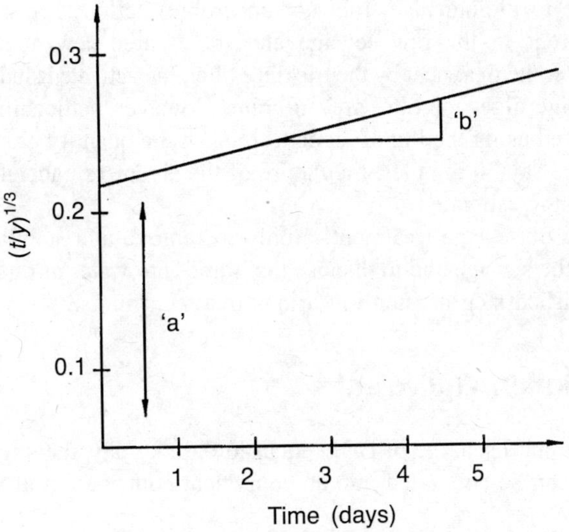

Fig. 4.18 Thomas method.

4.2.4 Chemical Oxygen Demand (COD)

The amount of oxygen needed for oxidising organic matter (including some inorganic solids) by chemical oxidation is known as chemical oxygen demand (COD). In a laboratory, COD is determined using a strong oxidising agent like potassium dichromate for oxidising organics (under aerobic conditions) into CO_2 and H_2O. The amount of oxygen consumed by the waste sample is its COD value. This depends upon a chemical reaction only not depending upon biological (seed) acclimation as in the case of BOD test. COD analysis in the laboratory will take few hours whereas BOD determination takes few days.

Most of the sewage treatment plants are biological units only. Hence the COD to BOD ratio must be less than 1.6 for effective decomposition of sewage in secondary treatment processes. If BOD/COD is less than 0.6 it may not be amenable for biological treatment.

4.2.5 Solids

Sewage contains different types of organics and inorganics. Organic solids are decomposed in treatment plants but most of the inorganic solids are in solution. Inorganic solids are also required to maintain optimum microbial activity. Some are floating solids forming a scum on the water surface, which affects reaeration. Oil and grease also clog and interfere in the treatment units.

Settling solids are of different sizes, shapes, densities and nature, with different settling velocities. In sewage treatment plants various units (like the grit chamber, the primary sedimentation tank, the detritus tank, the skimming tank, the secondary sedimentation tank) remove solids from the sewage. The sludge characteristics are changed due to biological (aerobic or anaerobic) action. The nature of the solids are different in the raw sewage and the treated sewage. The sludge characteristics after the treatment by the trickling filter, the activated sludge treatment process, the sludge digestion, etc. are different. However, settleability of solids can be easily tested using the Imhoff cone test. From the quantity of solids settled in a standard time say 30 minutes, the nature of the sludge is understood and the treatment efficiency estimated.

The purpose of sewage treatment is only to remove and stabilise the solid impurities from the sewage and to dispose the liquid into water or on land, safely. This is accomplished step by step in various treatment units.

4.2.6 Population Equivalent

Assuming per capita discharge of BOD equal to 0.08 kg/day, the strength of any industrial waste can be converted into an equivalent contribution of wastes from human resources.

$$\text{Population equivalent} = \frac{\text{BOD}_5 \text{ of industrial waste in kg/day}}{0.08 \text{ kg/day/person}}$$

This helps in charging the industry on a rational basis.

4.3 PRIMARY TREATMENT OF SEWAGE

4.3.1 Treatment Objectives

The tolerance limits for the sewage and the industrial effluents discharged into surface water sources are BOD_5 (20–30 mg/L); suspended solids (100 mg/L), nitrogen (50 mg/L), oil and grease (10 mg/L), phenol (1 mg/L), pH value (5.5 to 9) etc. The tolerance limits of waste water effluents to be discharged on land for irrigation purposes are BOD_5 (200 mg/L), pH (5.5 to 9), total dissolved salts (200 mg/L), oil and grease (30 mg/L), chlorides (600 mg/L), sulphates (1000 mg/L), etc. The sewage contains extremely harmful substances that are hazardous to health. Hence sewage treatment is essential before disposal. The purposes of sewage treatment are the following:

1. To prevent water borne diseases
2. To remove organic solids from sewage; to convert solids into stable products by biological decomposition
3. To dispose effectively and safely treated effluents
4. To safeguard water sources from pollution and
5. To prevent nuisance and offensive odours.

4.3.2 Schematic Outline

The primary treatment of sewage is carried out in various stages through screens, the grit chamber, the skimming tank, the primary sedimentation and the detritous tank. Secondary (biological) treatment consists of trickling filters, an activated sludge treatment process, the oxidation ponds, the sludge digester, the sludge drying beds, etc. An advanced treatment is given for the removal of specific characteristics from the sewage, as a special case. Chemical precipitation and chlorination are adopted only when they are required. Finally, the treated waste waters are disposed on land or water bodies.

The plan of treatment is shown in Figs. 4.19 and 4.20. The purification process in each unit is given below.

1. *Screen chamber* removes dead animals, branches of trees, logs of wood, rags and other coarse floating materials
2. *Grit chamber* removes heavy settleable inorganic solids such as sand
3. *Skimming tank* removes oil and grease
4. *Sedimentation tanks* are designed to separate solids from liquids under gravity and remove suspended settleable organic solids (about 60%) and BOD load of 30% associated with solids removal
5. *Trickling filters* are devices that depend on microorganisms for removing the settleable, non settleable, and dissolved organic solids (70–90%) and BOD load (65%–70%)
6. *Activated sludge treatment* method utilises additional oxygen and allows greater contact between waste water and microorganisms for the removal of organic load (80–90%) and BOD reduction (70%–80%)

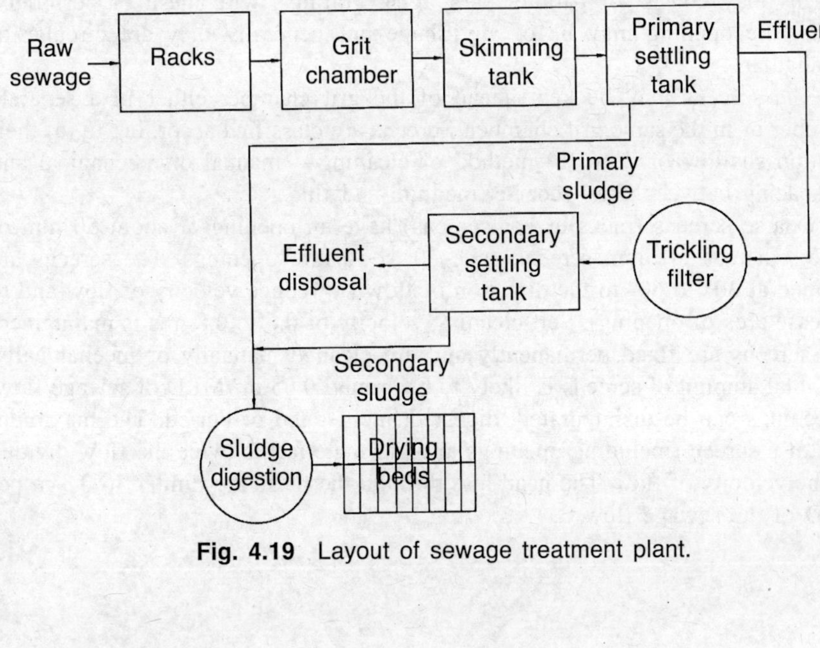

Fig. 4.19 Layout of sewage treatment plant.

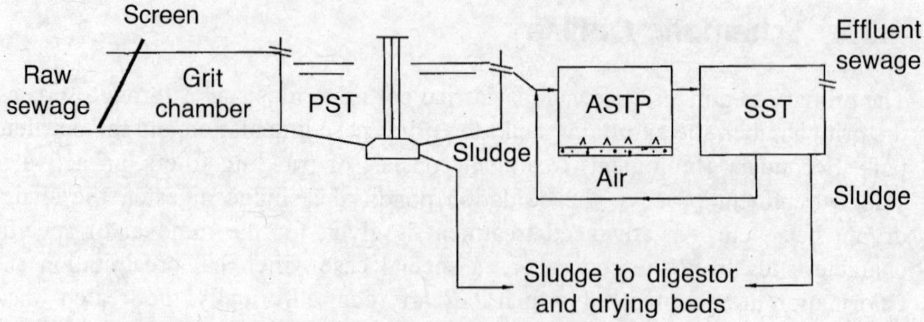

Fig. 4.20 Sewage treatment plant.

7. *Oxidation ponds* maintain aerobic conditions for the removal of organic matter and BOD (80–90%)
8. *Septic tanks* are individual units constructed to remove the solids (50%)

In general, waste waters (after purification in biological units and settling in secondary sedimentation tank) are disposed off ultimately in large bodies of water or on land. The sludge is collected from various units, digested, dried and utilised as manure or fuel.

4.3.3 Screens

A screen is the first unit employed in waste water treatment. Screens are provided to retain floating matter and coarse solids such as pieces of cloth, garbage, etc. so as to protect pumps and other units from clogging. The collected material is disposed off after cleaning (Fig. 4.21). A bypass channel is provided for the smooth passage of the sewage flow, in case of clogging of the screens. Screens may consist of parallel bars, rods, gratings, wire mesh or perforated plates, the opening may be of any shape but generally they are circular or rectangular.

The screens may be kept ahead of the grit chamber either in a separate chamber or in the same grit chamber. Screens are classified according to (a) their location shallow or deep (b) method of cleaning— manual or mechanical and (c) spacing between bars—coarse, medium and fine.

Coarse screens (racks or bar screens) have an opening of about 50 mm or more, whereas medium screens have 20 to 50 mm openings. The screens are inclined at 30° to 60° to the direction of flow (to reduce velocity of flow and to increase area of opening). Self cleaning velocity of 0.6 to 0.8 m/s is maintained. The screens are fixed permanently and are cleaned manually or mechanically. The total amount of screens is likely to be around 0.05 m^3/MLD of sewage flow. Screenings can be disintegrated, digested, incinerated or buried. The maximum area of a screen (including openings and bar material) is twice the flow divided by the velocity of flow. The head loss through the racks are limited to 0.3 m per MLD of the sewage flow.

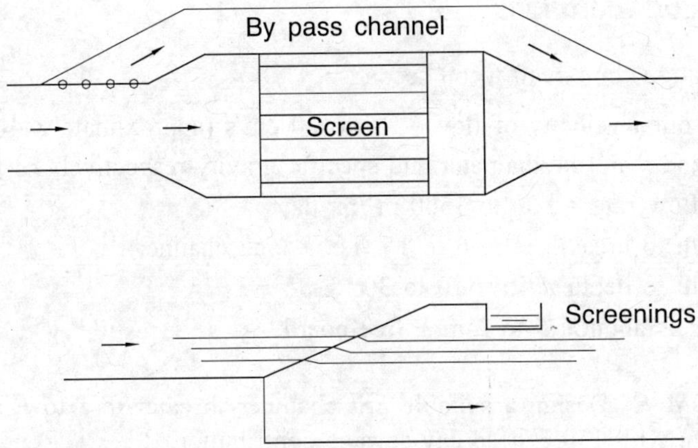

Fig. 4.21 Bar screen.

4.3.4 Grit Chamber

The grit includes sand, ash, cinder, egg shells, etc. of diameter less than 0.2 mm. Grit is non-putrescible and possesses a higher hydraulic subsidence value than organic solids. Hence, it is possible to separate the gritty material from organic solids by differential sedimentation in a grit chamber when the velocity of incoming waste water is reduced to 0.3 m/s. While organic matter will be in suspension (Fig. 4.22), the grit will separate which is washed before disposal.

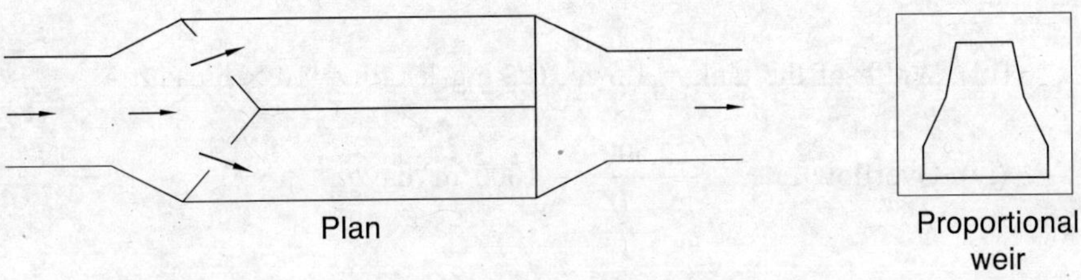

Fig. 4.22 Grit chamber.

The grit chamber is provided to remove the inorganic solids (specific gravity about 2.5, diameter less than 0.2 mm) like sand, ash and inert (non-putrescible) materials which may create operational problems. About 0.1 m^3/ML of sewage are likely to be removed. Velocity of flow (15 to 30 cm/s) is chosen such that grit only settles and other organic suspensions are carried through. Normally the quantity of sewage flow fluctuates and the velocity also varies. But in the grit chamber, constant velocity is maintained by designing a proportional weir outlet whose sectional area varies while discharge (Q) is proportional to the depth of flow (d), or

$$Q = 4.17(bd^{1/2}) \times d$$

For different values of depth, the value of breadth is obtained since the value of $bd^{1/2}$ is constant (which equals Q/d).

Design considerations

Detention time = 40 to 60 seconds

Horizontal velocity of flow = 15 to 30 cm/s (approximately $4\sqrt{g(s-1)d}$ where d and s are the diameter and specific gravity respectively of the particle)

Overflow rate = 500 to 1500 m^3/m^2/day

Length to breadth ratio 6 to 15 (i.e. a long channel)

Length to depth ratio = 10 to 30

Depth is about 1.5 to 2 m + free board

PROBLEM 6 Design a suitable grit chamber to cater to a town of 1.0 lakh population, with 150 L/head/day sewage contribution.

Solution

Maximum waste water flow is taken as $1.5 \times \dfrac{100{,}000}{1{,}000} \times 150$

i.e. 22,500 m^3/day = 0.23 m^3/s

(i) Assuming horizontal flow velocity 0.3 m/s;
length of the tank = velocity × detention time
∴ Length = 0.3 × 50 = 15 m

(ii) Assuming settling velocity = 0.02 m/s, depth of the tank = 0.02 × 50 = 1m.

(iii) Width of the tank = 1.5 m, (taking length = 10 × breadth)

(iv) Overflow rate = $\dfrac{22{,}500}{15 \times 1.5}$ = 1000 m^3/day/m^2

4.3.5 Primary Sedimentation Tank

The simplest method of removing the suspended impurities is by plain sedimentation. Waste water is allowed to stand quiescent or move very slowly through the natural or artificial basins until the suspended impurities settle to the bottom and the relatively clear water is drawn from the top portion (Fig. 4.23). The factors that influence the sedimentation include (a) characteristics of solids like size of particles, the sp. gravity of the settling particles, the concentration of the suspended matter (b) characteristics of the liquid like the temperature, viscosity, specific gravity (c) physical characteristics of settling basin like the period of detention, the shape of basin, the velocity and length of flow through the basin. They are provided for removing organic settleable solids (specific gravity 1 to 1.2) by gravitational settling. The velocity of flow, detention time and overflow rate control the performance of the settling unit.

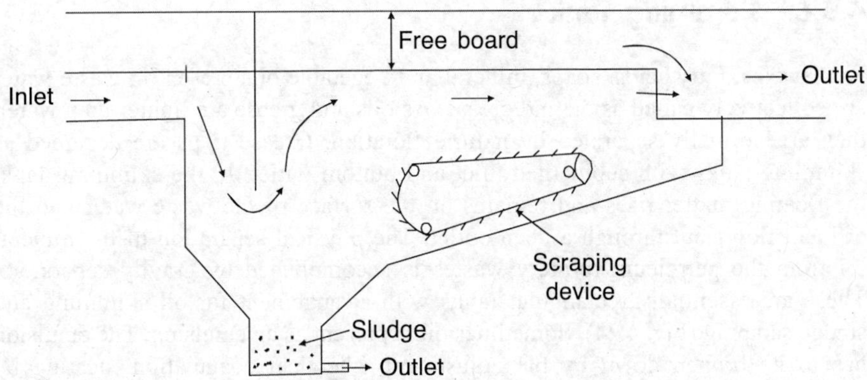

Fig. 4.23 Sedimentation tank.

Design considerations

Detention time = 2–3 hours (longer detention causes putrefaction of sewage)

Overflow rate = 30–50 m^3/m^2/day (less rate more efficient removal of BOD)

Velocity of flow = 15–30 m/minute

Depth of the tank = 3–4 m, width of the tank = 5 to 8 m.

Weir loading = about 300 m^3/day/meter length

Tank is provided with a bottom slope for collecting sludge effectively with the help of a mechanical cleaning device.

PROBLEM 7 Design a primary sedimentation tank for treating 22,500 m^3/day of sewage.

Solution

(i) Assuming 2 hours detention time,

$$\text{volumetric capacity of the tank} = 22{,}500 \times \frac{2}{24} = 1875 \text{ m}^3$$

(ii) Length of the tank = Velocity of flow × detention time = 0.3 × 2 × 60 = 36 m

(iii) Provide two tanks with an effective depth of 3 m; 0.5 m free board and 0.5 m for sludge storage

(iv) $$\text{Width of the tank} = \frac{1875}{2 \times 3 \times 36} = 9 \text{ m (approx.)}$$

(v) $$\text{Surface area loading rate} = \frac{22{,}500}{2 \times 9 \times 36} = 35 \text{ m}^3/\text{m}^2/\text{day}$$

(vi) $$\text{Weir length required} = \frac{22{,}500 \text{ m}^3/\text{day/m}}{2 \times 300 \text{ m}^3/\text{day}} = 37.5 \text{ m}$$

4.3.6 Skimming Tank

Fats, waxes, fatty acids, soaps, mineral and vegetable oils present in waste water are collectively called as oil and grease. As oils and grease are lighter than water, they are normally separated by natural flotation. Grease traps are designed as skimming tanks with submerged inlet and bottom outlet. In the skimming tank, the floating matter rises and remains on the surface of the waste water and the effluent flows out through a deep outlet. The physical separation of the mineral oil from the petroleum refinery wastes is accomplished by gravity separators. These are essentially rectangular tanks with arrangement for oil skimming and sludge scraping (Fig. 4.24). Sometimes oil is present as an emulsion. The emulsion has to be broken down by pH adjustment, chemical coagulation, heating by adding salt or by ultrafiltration.

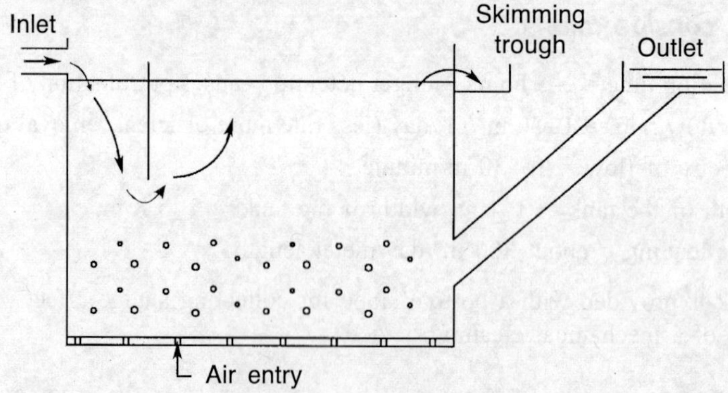

Fig. 4.24 Skimming tank.

For effective removal, compressed air is sent through diffuser plates located at the bottom of the tank. Sometimes, a skimming trough may be provided in a sedimentation tank to collect the floating matter from sewage. The surface area of the tank is determined using an overflow rate of 30–40 m^3/m^2/day. Aeration easily separates the scum from the suspended solids, because of gentle agitation.

4.3.7 Other Units

1. *Detritus tanks* (settling tanks) are provided to remove the fine organic solids.
2. *Chemical precipitation* using the coagulants ($FeSO_4$, $FeCl_3$) is not generally practised in sewage treatment. No purpose is served by collecting voluminous sludge without destroying its putrescible nature. However, if biological treatment is not provided and the suspended solids are less in sewage, chemical precipitation and chlorination is done as a method of final disposal.

3. All these units viz. the screens, the grit chamber, the sedimentation tank are classified under primary treatment of the sewage. In practice, no biological action takes place in these units.

4.4 BIOLOGICAL TREATMENT

4.4.1 General

Aerobic treatment processes utilise a mixed population of microorganisms which convert the dissolved organic material into new cellular material and CO_2 in the treatment system. This mixed microbial population, referred to as heterogenous population, continuously changes in nature in response to variation in the environmental conditions and influent waste characteristics. All secondary treatment plant units provide opportunity of contact between the bacteria, the organic matter, and the oxygen for aerobic reactions. In anaerobic conditions, treatment units do not provide air supply.

The following points are of interest.

1. The substrate (organic matter) is converted into stable compounds due to biological actions based on the catalytic activities of enzymes. Production of enzymes controls the rate of decomposition.
2. In chemical reactions with a single reactant, the removal rate (kinetics) dc/dt depends upon the order of reaction, such as

$$\frac{dc}{dt} = -K \quad \text{(zero order reaction)}$$

$$\frac{dc}{dt} = -KC \quad \text{(first order reaction)}$$

$$\frac{dc}{dt} = -K(C)^2 \quad \text{(second order reaction).}$$

   where K = reaction constant and C = concentration level.
   But in sewage, the chemical reactions take place with mixed culture of microorganisms in the multi substrate media.
3. Temperature affects the rate of reaction, given by the formula

$$K_T = K_{20}\,(1.05)^{(T-20)}$$

   where K_T and K_{20} are reaction rates at temperatures of T°C and 20°C.
4. The growth of microorganisms depends upon time, availability of food and favourable conditions. Initially there will be a slow-adaption-lag phase, and later on growth rate accelerates. When the substrate decreases, microorganisms also have a declining rate which leads to an endogenous (death) rate. In a continuous process, new cells are always produced and useless biomass is wasted out from the circulation again.

5. Microorganisms act at an optimum range of 6–8 pH. Two distinct temperature ranges exist for biological action as a mesophilic range of 25°–45°C and a thermophilic range of 55°–65°C.

The aerobic bacteria utilise dissolved oxygen. Some bacteria (anaerobic) also gains energy from nitrates, sulphates, etc. for growth and multiplication. Autotrophic bacteria use carbon source and oxidise inorganic matter for their energy. The heterotrophic bacteria take energy from organic matter. The saprophytic bacteria live on dead or decaying organic matter. Many different types of organisms including bacteria, fungi, protozoa, rotifers and other forms of life can be found in the aerobic treatment system at any particular time. Of the various forms present, bacteria are the most important organisms as they represent the major functional group in the biological system. Biological removal of degradable organics involves a sequence of steps including mass transfer, adsorption, absorption and the biochemical enzymatic reactions.

The stabilisation of organic substances by the microorganisms in a natural environment or in a controlled environment of biological treatment process is accomplished by two distinct metabolic processes: (i) respiration and (ii) synthesis. Respiration is a microbiological process in which a portion of the organic matter present in waste water is oxidised by enzymes (biocatalysis) to liberate energy. Dissolved oxygen acts as the final hydrogen-acceptor for oxidation with end products CO_2, H_2O and NH_3. The energy derived from respiration is utilised to synthesize new microbial cells enzymatically.

Virus, bacteria, fungi (multicellular aerobes), algae (produce oxygen), protozoa (microscopic animals), rotifers (aerobic and heterotrophic animals), and crustaceans are present in waste waters undergoing decomposition. The rotifers and crustaceans indicate (low organic load and high dissolved oxygen) a reasonably good quality of effluent.

In a complete mix (continuous flow and thorough mixing) type of reactors, microorganisms and organic matter are in suspension stage as in the activated sludge process and in attached growth stage, in a fixed trickling filter bed. If the movement is in one direction only (along the length of the tank), it is known as a plug flow reactor.

The main design considerations are: food to microorganism ratio, mean cell residence time, and specific substrate utilisation rate. Moreover, the following points also should be considered while designing a reactor.

- If the food to microorganism ratio is low, the BOD removal will be high.
- The organic loading is the BOD applied per day to the aeration tank per unit volume of the tank.
- The sludge age indicates the retention time of biological solids in the system. As solids are recycled from the secondary sedimentation tanks, this sludge age is likely to be more in the aeration tank.

Recirculation

In the high rate trickling filter and activated sludge process, recirculation of (aerobically) the treated sewage solid sludge with aerobic bacteria is essential for obtaining the required efficiency of treatment (Fig. 4.25). Recirculation gives more opportunity of contact between the organic solids and the biological slimes. Uniform flow rates can be obtained inspite of fluctuations in the sewage flow. As the sewage remains fresh, odours are reduced.

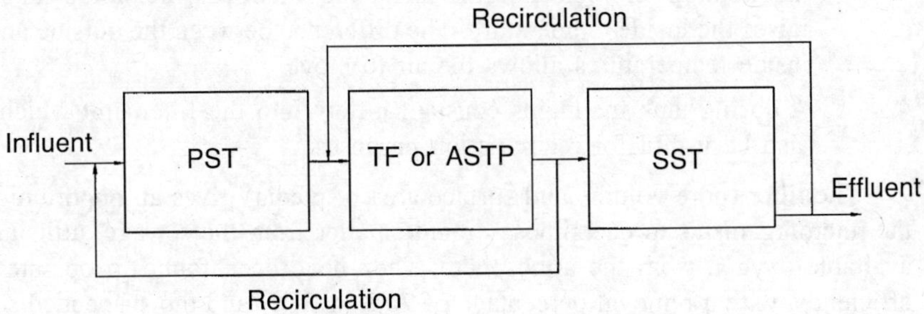

Fig. 4.25 Recirculation.

Recirculation factor $(F) = (1 + R/I) \div (1 + 0.1\, R/I)^2$, where R = the recirculated sewage and I = the raw sewage. The value of F is 1, without recirculation in a standard rate trickling filter.

4.4.2 Trickling Filters

In these filters, (Fig. 4.26) sewage trickles and percolates through the filtering media. The essential components are discussed here.

1. Rotary distributors are fitted with two or four arms, supported at the centre. As the arms move, sewage passes into the bed before receiving another load, uniformly on the entire surface area.

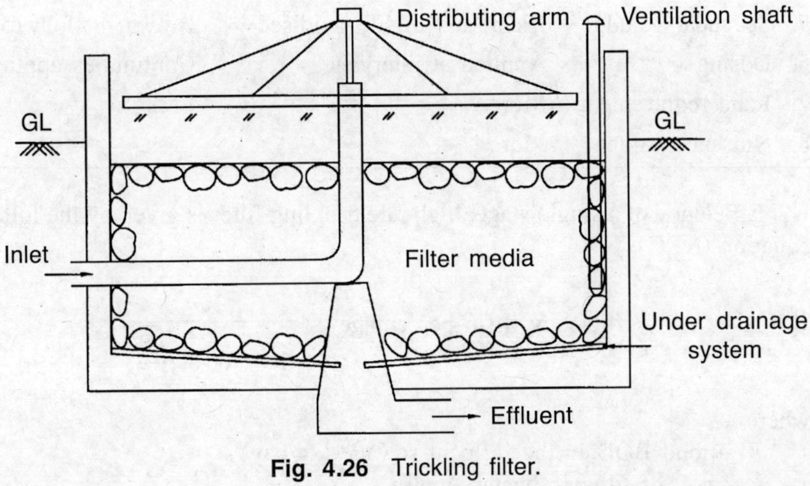

Fig. 4.26 Trickling filter.

2. The filtering media consists of hard (granite) coarse broken stones of 30 to 60 mm size, and about 2-m depth.
3. The vitrified clay blocks are used as under-drains to collect the filtered sewage. The floor bottom should have uniform slope for easy collection.
4. A ventilation shaft is also provided if the air flow is not sufficient for maintaining aerobic conditions inside the filter bed. The biological action raises the inside temperature. The difference between the outside and the inside temperatures allows the air to move.
5. A dosing tank maintains constant inflow into the filter unit which will also be useful for recirculation purposes.

The filter (pore volume and surface area of media) gives an opportunity for the microorganisms to decompose organic matter from the sewage, utilising the available oxygen from the atmosphere. They have been found to operate very efficiently, with a removal percentage of 70% in BOD and the suspended solids.

Trickling filters are not very sensitive towards input fluctuations. Normal filters have the disadvantage of attracting flies, and besides odour nuisance and clogging troubles due to heavy microbial growth. As the rate of loading is less, more land area is required. As recirculation of fresh seeded solids improves the efficiency, the trickling filters are operated under high rates.

The low rate and the high rate filters are compared as follows:

Characteristics	*Standard rate trickling filter*	*High rate trickling filter*
1. Hydraulic loading (m^3/m^2/day)	1–4	10–40
2. Organic loading (kg/m^3/day)	0.08–0.3	0.3–1.5
3. Recirculation	Not required	Essentially provided
4. Effluent	Highly nitrified and stabilised	Not fully nitrified
5. Secondary sludge	Black and highly oxidised	Brown, not fully oxidised
6. Dosing	Applied at intervals	Continuous application
7. Land requirement	More	Less
8. Nuisance value	More	Less

Efficiency of a single stage, high rate trickling filter is given by the following equation (NRC).

$$\text{efficiency } \eta \text{ (in percentage)} = \frac{100}{1 + 0.0044\sqrt{\dfrac{W}{VF}}}$$

where

W = total BOD in the influent sewage (kg/day)

V = vol. of filter in hectare-metre

F = recirculation factor

PROBLEM 8 Determine the size of a high rate trickling filter from the following data:

(i) Sewage flow = 5.0 MLD
(ii) BOD of sewage input to filter at 170 mg/L = 850 kg/day
(iii) Efficiency of filter = 80%
(iv) F = recirculation factor = 1.85

Solution

Using the formula, we have

$$\text{Efficiency of filter} = \frac{100}{1 + 0.0044\sqrt{\dfrac{\text{BOD applied}}{VF}}}$$

where V = volume of filter in hectare metre

or,
$$80 = \frac{100}{1 + 0.0044\sqrt{\dfrac{850}{V \times 1.85}}}$$

$\therefore$ Volume of the filter = 1440 m^3

Provide two filters of 25-m diameter and depth 1.75 m

PROBLEM 9 Design a standard rate trickling filter, assuming a hydraulic loading of 2 m^3/m^2/day and an organic loading of 0.16 kg/m^3/day to treat 5.0 MLD flow and 850 kg/day of BOD_5 into the filter unit.

Solution

(i) Surface area of the filter required $= \dfrac{5\text{ MLD}}{2\text{ m}^3/\text{m}^2/\text{day}} = \dfrac{5000}{2} = 2500\text{ m}^2$

(ii) Volume of the filter $= \dfrac{850\text{ kg/day}}{0.16\text{ kg/m}^3/\text{day}} = 5310\text{ m}^3$

Provide two filters having diameter = 40 m and depth = 1.5 m

4.4.3 Activated Sludge Treatment

Sewage from a sedimentation tank enters into an aeration tank. Here 20 to 30% of active sludge is mixed. The mixture is aerated and mixed in the tank for about 4 to 8 hours residence time. The microorganisms oxidise organic matter, in the presence of abundant quantity of oxygen in the aeration tank. Sewage is allowed to settle in a secondary sedimentation tank. This settled sludge has undergone aeration and has active microorganisms. So, some portion of this active sludge is recirculated into the aeration tank, where the mixed liquor containing the suspended solids is under continuous agitation. The excess sludge is disposed off from the secondary sedimentation tank.

A typical activated sludge plant flow diagram is shown in Fig. 4.27 and an aeration unit is shown in Fig. 4.28.

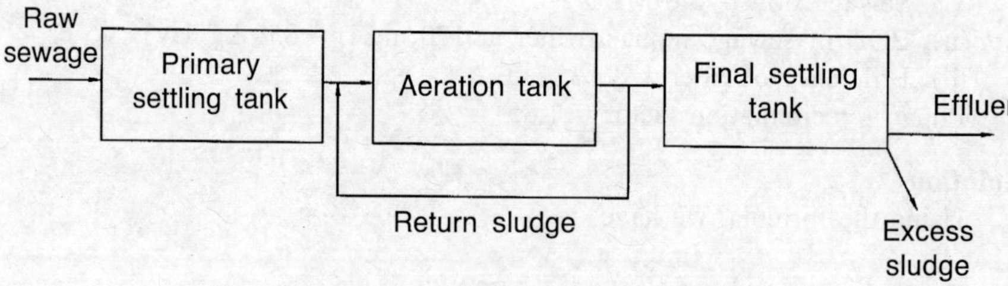

Fig. 4.27 Activated sludge process.

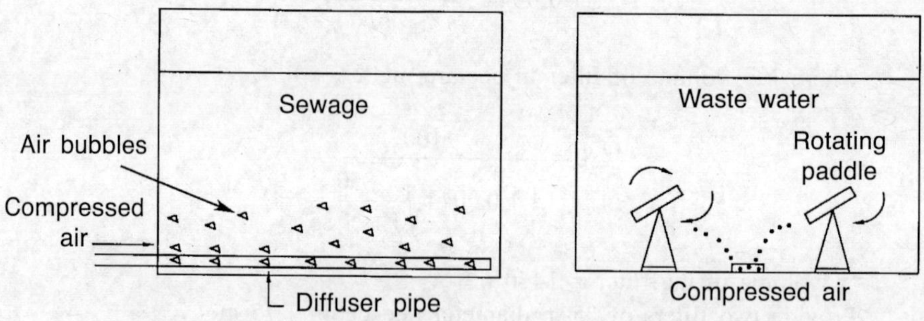

(a) Diffuser aeration system (b) Mechanical aeration system

Fig. 4.28 Aeration units of ASTP.

In an ASTP unit, skilled operation is needed to maintain proper aeration, mixing, food to microorganism ratio and recirculation of sludge.

There are two types of aeration, diffused aeration and mechanical aeration. In diffused aeration system, compressed air bubbles are introduced through submerged nozzles or diffusers, near the bottom of the tank. In mechanical aerators, in addition to air supply, mixed liquor is under agitation by the mechanical (paddle) aerators.

The main disadvantage of this ASTP, is bulking (increasing the volume for unit weight) of sludge. Sometimes sludge blanket formation may occur at the top surface level in the tank.

Design considerations

1. Aeration period—4 to 8 hours, detention time (h) = $\frac{\text{BOD}}{20} - 1$ or $\frac{(\text{BOD})^{3/4}}{10}$
2. BOD loading—300 to 700 g/m^3 of tank volume
3. Amount of O_2 required—0.8–1.0 kg/kg BOD removal or amount of air required—50 to 100 m^3/kg of BOD removal

4. Mixed liquor suspended solids (MLSS)—1500–3000 mg/L
5. Food/microorganism ratio—0.4–0.2

$$\text{organic loading rate} = \frac{\text{BOD applied}}{\text{MLSS}}$$

6. Depth of tank—3 to 5 m, width 5–10 m and length 40–50 m.

The hydraulic retention time may be defined as the rate of application of sewage by volume in the aeration tank. The mean cell residence time is the sludge age represented by MLSS (mixed liquor suspended solids) per unit of suspended solids (SS) lost in the effluent system.

PROBLEM 10 Design a conventional activated sludge treatment plant from the given data:

Waste water flow = 30,000 m^3/day
Aeration period = 6 h
Influent BOD = 200 mg/L, MLSS = 2500 mg/L
F/M = 0.3 (approximately)

Solution

$$\text{Volume of the aeration tank} = Q \times T = 30{,}000 \times \frac{6}{24} = 7500 \text{ m}^3$$

$$\text{F/M ratio} = \left(\frac{30{,}000 \times 200}{1000}\right) \div \left(\frac{7500 \times 2500}{1000}\right) = 0.32$$

Assume tank dimensions as depth = 3–4 m

width = 20 m and length = 60 m, two tanks.

Air required at the rate of 100 m^3/day of kg of BOD removed

$$= \frac{30{,}000 \times (200 - 25)}{1000 \times 1000} \times 100 = 525 \text{ m}^3\text{/day}$$

Modifications of the process

In a conventional process, the supply of oxygen is uniform with the time period and the length of the channel. But technically the oxygen demand and the F/M ratio vary. Most of the process modifications are based on air supply to the tank.

(a) Tapered aeration ensures higher air supply near the inlet portion of the tank and less along the flow path. So diffuser plates or surface aerators are spaced accordingly.

(b) Step aeration involves in providing inlets for the sewage along the length of the tank. This ensures better utilisation of oxygen as recirculated sewage reaches at various lengths of the aeration tank. In this case, air supply is uniform throughout the tank.

(c) In the biosorption process, an additional contact basin is provided for mixing the sewage and the returned sludge for an hour. The effluent is settled and transferred to a sludge aeration tank for a period of 3–5 hours and recirculated again.

(d) In a complete mix type, a number of inlets are provided along the length of the channel separately for both incoming sewage and return sludge. This ensures thorough mixing and optimum oxygen utilisation as sewage and return sludge get mixed up uniformly inside the tank volume.

(e) In a modified aeration process (where higher efficiencies are not required), the aeration periods and the recirculation rates are reduced, whereas F/M ratio and hydraulic loading are kept high.

(f) Extended aeration system is similar to complete mix process operating under endogenous growth phase. Oxidation ditch is an example of this method for treating sewage with very high degree of efficiency, from small communities. In this process, aeration period extends upto 24 hours with low organic loading and F/M ratio (0.1–0.15) and high MLSS concentration (4000 mg/L).

4.4.4 Oxidation Ponds

These are simple, low cost holding basins 1–2 m deep taking oxygen (air) from the atmosphere. The algae supply the additional oxygen through the process of photosynthesis in shallow ponds, under aerobic conditions. If the depth is 3–5 m, decomposition takes place under anaerobic conditions at lower depths. These ponds may accept higher organic loadings.

In the facultative ponds, the upper layers are under aerobic composition and lower layers are under anaerobic conditions. In the middle portion, the algae may release oxygen to the upper layers. From the settled organic load (benthal deposits), methane, ammonia, carbon dioxide, etc. may be released. These ponds may be kept in series for improved efficiency.

The essential feature of the oxidation pond is that bacteria and algae live together (symbiosis) for mutual benefits. CO_2 and the nutrients are used by algae (green) and O_2 is released in daylight.

$$CO_2 + H_2O \rightarrow CH_2O + O_2$$

This photosynthetic release of oxygen depends upon the latitude of the place. Yield of photosynthetic oxygen is likely to be about 250 to 150 kg/hectare/day for places having 20° to 36° N latitude. In shallow ponds, wind force naturally mixes the contents of the pond and surface aeration takes place. Bacteria oxidise organic matter present in the sewage.

Various processes like sedimentation, oxidation, digestion, photosynthesis, gas exchange, evaporation and seepage take place in an oxidation pond (Fig. 4.29(i)).

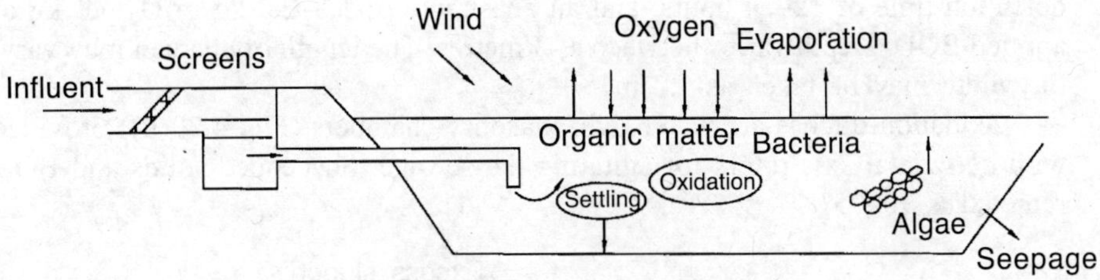

Fig. 4.29(i) Oxidation pond.

Design considerations

(1) Detention period is usually 10–40 days depending upon latitude of the place (temperature and sunlight availability).

(2) Organic loading may be taken from 300–150 kg/hectare/day depending upon the climate and location of the place.

(3) The pond depth is taken as 1–2 m. Rectangular ponds of length to breadth ratio of 2 may be considered.

As mosquito and odour nuisance is expected, these ponds are kept away from residential areas, considering the wind direction.

PROBLEM 11 Design an oxidation pond given that

Flow = 1 MLD, BOD = 150 mg/L, Depth of pond = 1.5 m.

Yield of photosynthetic oxygen = 250 kg/hectare/day

Solution

(i) $$250 = 15\,\frac{\text{depth of pond (m)} \times \text{influent waste water BOD (kg/day)}}{\text{hydraulic detention period (days)}}$$

$$\therefore \text{Detention time} = \frac{15 \times 1.5 \times 150}{250} = 13.5 \text{ days}$$

$$\text{Surface area of the pond} = \frac{1000(\text{m}^3/\text{day}) \times 13.5(\text{day})}{1.5(\text{m})} = 9000 \text{ m}^2$$

Provide 4 ponds of 35 m × 75 m

or (ii) Assuming an organic load of 150 kg/hectare/day

$$\text{Pond area required} = \frac{150 \text{ (kg/day)}}{150 \text{ kg/hectare/day}} = 1 \text{ hectare}$$

4.4.5 Oxidation Ditch

This is an activated sludge treatment plant under extended aeration type, with detention time of 12–24 hours, and an air supply of 1.5–2.0 kg of O_2 per kg of applied BOD. Depth varies between 1–3 metres. The length of channel may vary; the width may be taken as 1.2 m.

Oxidation ditches are water tight masonry chambers (Fig. 4.29(ii)) provided with horizontal axis rotors for agitating the sewage. Suspended solids and BOD removal is 90–95%.

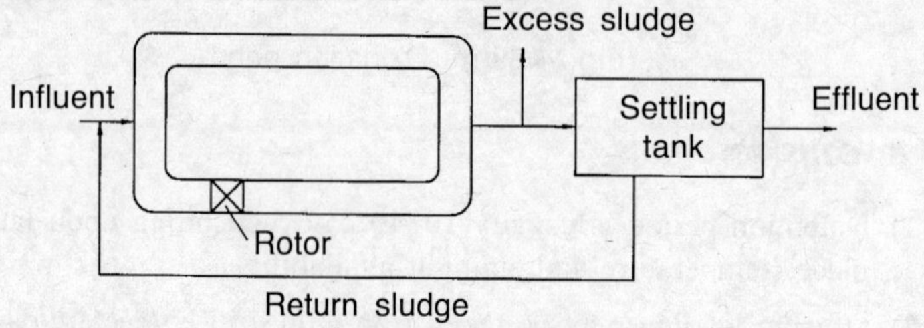

Fig. 4.29(ii) Oxidation ditch.

4.4.6 Rotating Biological Disc Contractors (RBC)

For the smaller installations, the microbial film is allowed to grow on the surface area of PVC discs, rotating (partially submerged) in the direction of the flow as shown in Fig. 4.30. These discs (2 m diameter and 100 mm thick) absorb (attach) organic load when submerged and take oxygen when exposed to the atmosphere. Substrate utilisation is a continuous process. They may have a hydraulic loading of 0.1 m^3/m^2 of surface area/day and organic loading of 50g BOD/m^2/day.

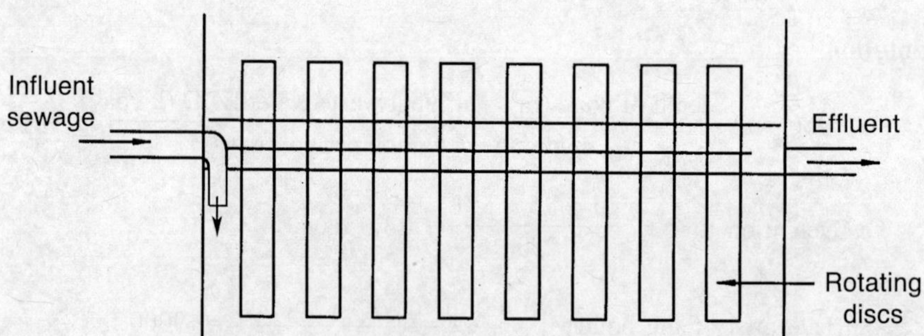

Fig. 4.30 R.B.C.

4.4.7 Secondary Sedimentation Tank

After the biological oxidation and nitrification, the suspended solids settle quickly and more completely in a secondary sedimentation tank. The settled portion is sent to a digestion tank for further treatment.

Detention time of 1–2 hours and an overflow rate of 20–50 m^3/m^2/day may be taken as design criteria.

4.5 SLUDGE TREATMENT AND DISPOSAL

The anaerobic treatment technologies are used for the effective treatment of municipal waste waters, sludge and industrial waste waters. This technology is particularly attractive as the energy required for operating the process is minimal compared to the energy required for aerobic processes. In addition, energy for on-site use is produced in the form of methane gas which can be converted into mechanical or electrical energy for use at the treatment plant to operate aeration blowers and/or pumps. Many improvements have been made in recent years in both fundamental and practical aspects of anaerobic treatment. The high rate digestor with mixing and temperature control, anaerobic filter, suspended and attached growth reactors, up-flow anaerobic sludge blanket (UASB) reactors and fluidised bed reactors have been developed and are being used.

4.5.1 Sludge Characteristics

The main objective of primary sewage treatment is to separate solids from waste water. The solids (sludge) in the form of a slurry contains putrescible substances which decompose further. The moisture content depends upon the type of treatment. But after sludge treatment, dewatering is easy. The sludge before the treatment can be distinguished from the sludge after the treatment. Digested sludge dewaters quickly on drying beds. Sludge volume index (SVI) indicates the nature of physical state of solids in a treatment system. It is the volume settled in millilitre by one gram of solids in 30 minutes settling time. Sludge with 90% moisture content is half in quantity to the sludge with 95% moisture. SVI is an operational parameter, lower the value, better the process of sludge settling.

4.5.2 Digestion of Sludge

Sludge is digested anaerobically, converting the waste solids into carbon dioxide and methane. Sludge volume is reduced as gases escape and liquefaction also takes place. Well digested sludge contains nitrogen and phosphates for use as a fertiliser. Methane gas is used as a fuel.

Mechanism of digestion

The acid forming bacteria convert complex organic substances into simple organic acids. This acid production stage lowers the pH value. Methane forming bacteria converts the acids into methane and carbon dioxide. Decomposition of carbohydrates, fats, proteins will take place in that order. Since carbohydrates are digested easily, bacteria decompose them first. As the first stage of digestion nears completion (in 10–15 days), volatile organic acids and nitrogenous compounds are converted into H_2S, CH_4, NH_3, CO_2. Foaming occurs and the pH value rises also a little indicating the acid regression stage. More resistant proteins and organics are converted in the final stage resulting in an alkaline nature of the sludge. In this alkaline fermentation stage CH_4 and CO_2 are evolved and BOD of the sludge is reduced.

$$4H_2 + CO_2 \rightarrow CH_4 + 2H_2O$$

Water separates out from the well digested sludge. This process of stabilising organic matter under anaerobic decomposition is known as sludge digestion. Different types of anaerobic reactors with low hydraulic retention time and high solids retention time like contact reactors are in use. Mixing the return sludge and maintaining the temperature reduce the detention time in digestors.

Rate of digestion is more at higher temperatures. The relationship between the temperature and the digestion period is shown in Fig. 4.31. Normally digestion is carried under mesophilic temperature range of about 30°C and the corresponding digestion period is about 30 days. If required, the temperature may be controlled inside with external heating devices. Methane forming bacteria are very active in a pH range of 6–8, and alkaline conditions are favourable for digestion. Proper seeding of a digested sludge is beneficial from operational point of view. Mixing of raw sludge and digested sludge improves the efficiency of a digestor. This may be achieved by providing suitable inlets and outlets.

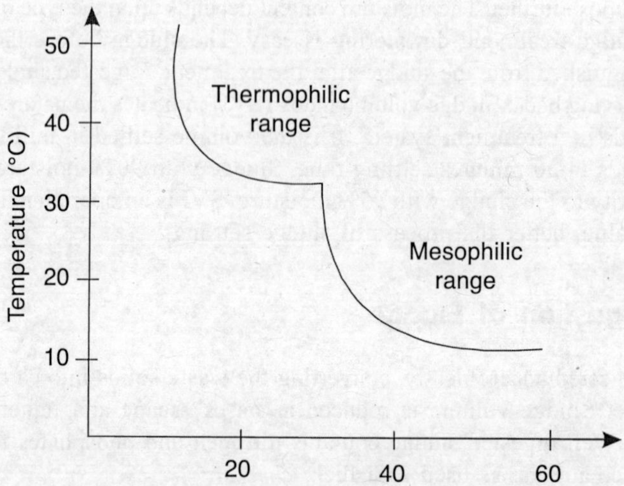

Fig. 4.31 Effect of temperature on digestion period.

Sludge digestion tank

The digestion tank (Fig. 4.32) (about 10–20 m diameter and 6-8 m depth) is provided with outlets and devices for

(a) the collection of gases

(b) the collection of supernant liquor and

(c) the collection of digested sludge

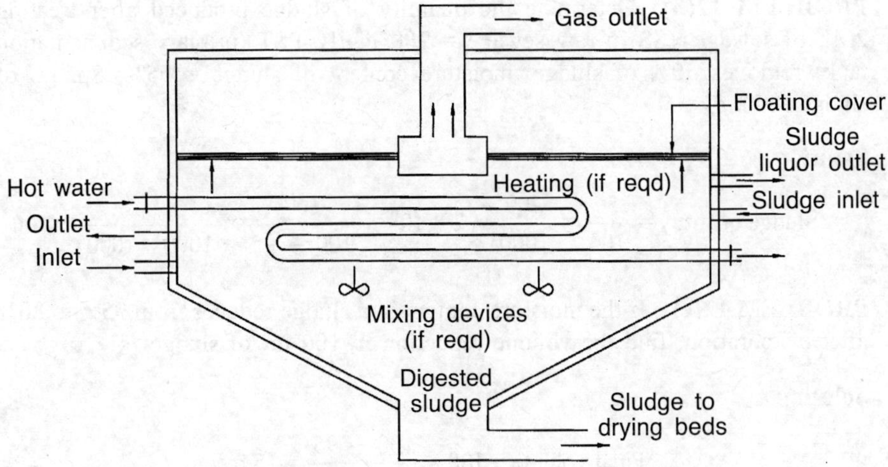

Fig. 4.32 Sludge digestion tank.

(d) breaking the scum

(e) scraping the sludge

(f) mixing the contents

It may have fixed roof with separate gas holding device or a floating type of roof. 0.016 m^3/capita (1m^3/kg of volatile solids) is the amount of gas production. Volume of a digestion tank may be calculated at 0.033 to 0.060 m^3/capita or from the relation

$$V = [V_f - \frac{2}{3}(V_f - V_d)]\, t_1 + V_d t_2$$

where

V = Volume of the digestor (m^3)
V_f = Volume of fresh sludge (m^3/day)
V_d = Volume of digested sludge (m^3/day)
t_1 = digestion period in days
t_2 = period of digested sludge storage

PROBLEM 12(a) Design a sludge digestor for a town with a population of 1.0 lakh, given that $V_f = 2.1$ m^3/day/1000 persons, $V_d = 0.8$ m^3/day/1000 persons, digestion period, $t = 30$ days and temperature = 27°C.

Solution

Volume of digestor per 1000 persons = $[V_f - 2/3(V_f - V_d)]\, t$

$= [2.1 - 2/3 \times 1.3] \times 30 = 37$ m^3

Provide 60 m^3 volume per 1000 persons, taking into consideration monsoon storage.

Therefore, Volume of the digestor = $\frac{60}{1000} \times 100{,}000 = 6000$ m^3

Provide 2 numbers of 20-m diameter and 10-m depth tanks.

PROBLEM 12(b) Determine the quantity of sludge produced after treating 2ML of sewage. (SS in raw sewage = 200 mg/L; PST (primary sedimentation tank) removes 50% of sludge; moisture content of sludge = 95%; Sp. gr. of sludge = 1).

Solution

$$\text{Sludge quantity} = \frac{50}{100} \times \frac{200}{1000} \times 2 \times 10^6 \times \frac{100}{100 - 95} \times \frac{1}{1000 \times 1000} = 4 \text{ m}^3$$

PROBLEM 12(c) If the moisture content in a sludge reduces from 95% to 80% after evaporation, find the volume reduction in 100 m^3 of sludge.

Solution

$$\text{Final volume} = 100 \times \frac{100 - 95}{100 - 80} = 25 \text{ m}^3$$

4.5.3 Sludge Drying Beds

On drying beds (Fig. 4.33), the water evaporates and percolates, leaving the solids to dry in 10–20 day's time. 0.1–0.2 m^2 per capita or 50–100 kg/m^2/year of dry solids may be sufficient for bed area. Sufficient number of extra drying beds are provided for alternate feeding or as standby.

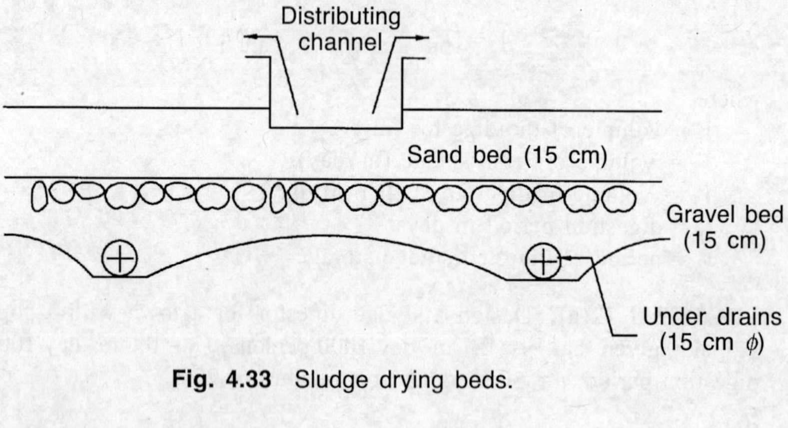

Fig. 4.33 Sludge drying beds.

4.5.4 Sludge Conditioning

Depending upon the need, the following operations are performed on the sludge before its disposal:

1. Chemical precipitation with coagulants (iron salts) with or without lime treatment
2. Disinfection to kill the pathogenic organisms by lime or chlorine

3. Heat treatment
4. Elutriation or washing of sludge with water to remove chemicals interfering with dewatering or precipitation
5. Thickening the sludge by gravity or centrifugal action
6. Vacuum filtration to dewater sludge
7. Centrifugation
8. Incineration to destroy organic matter.

Finally the solid sludge can be used as a fertiliser effectively.

4.5.5 Septic Tank

In a septic tank (Fig. 4.34), sewage and sludge are detained for 12 to 24 hours in a closed water tight chamber. Sedimentation and digestion are carried out in this septic tank under anaerobic activity. Foul gases escape through a vent shaft. The digested sludge is periodically removed once in a year or two, and disposed. The effluent from the septic tank is highly offensive (200 mg/L organic solids and 100 mg/L BOD) and hence disposed into ground through soak pits (Fig. 4.35) or dispersion trenches. Septic tanks should be commissioned with well-digested sludge. Septic tanks are very useful for individual houses, hostels and small groups of populations.

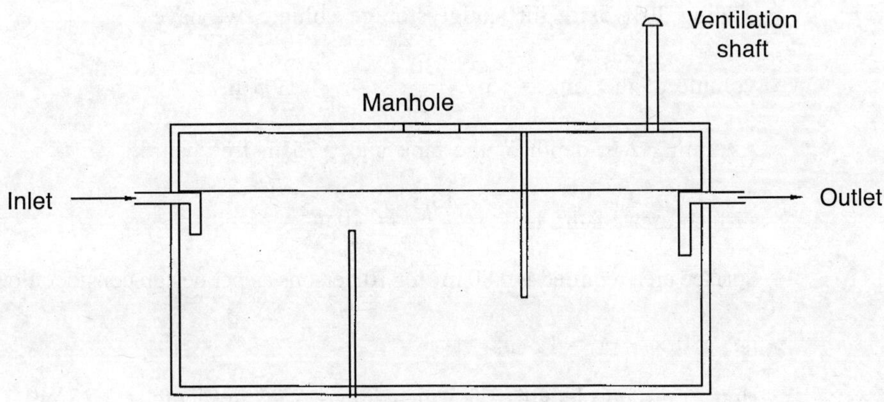

Fig. 4.34 Septic tank.

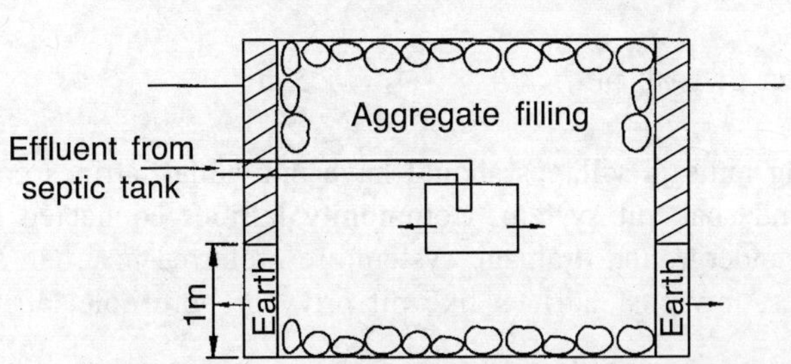

Fig. 4.35 Soak pit.

Design considerations

Detention period—12–24 hours

Surface area required is taken at 0.8–0.9 m^2 for 10 persons.

Depth 2–3 m.

Length to breadth ratio 3 : 1

Minimum capacity of tank 1.5–2.5 m^3 for 10 persons.

Free board 0.25 m.

The outlet invert level should be kept few centimetres below the invert level of the inlet. A soak pit is designed on the basis of percolating capacity of filtering media, as 1–1.5 $m^3/day/m^3$ of volume or soil absorption capacity may be taken as 50 to 200 $L/m^2/day$.

PROBLEM 13 Obtain the dimensions of a septic tank for a hostel (150 persons) at 100 L/head/day sewage contribution.

Solution

(i) Assuming a detention period of 24 hours, we have

$$\text{Capacity of the septic tank} = \frac{150 \times 100}{1000} \times \frac{24}{24} = 15 \text{ m}^3$$

Adding 30% extra for sludge storage volume, we have

$$\text{Volume of the tank} = 15 + \frac{30}{100} \times 15 = 19.5 \text{ m}^3$$

Assuming 2-m depth of the tank and 0.25 m free board,

$$\text{surface area of the tank} = \frac{19.5}{2} \simeq 10 \text{ m}^2$$

Surface area required = 0.80 m^2 for 10 persons as per design considerations,

$$\text{i.e. } 150 \times \frac{0.8}{10} = 12 \text{ m}^2$$

Hence, provide length = 5.5 m, width = 2.25 m, depth = 2.25 m

(ii) Assuming 3 m depth, surface area of soak pit = 15/3 = 5 m^2

$$\text{Diameter of soak pit} = \sqrt{4} \times \frac{5}{\pi} \simeq 2.5 \text{ m}$$

All dwelling units in villages should have individual latrines connected to a septic tank and soak pit system. Community latrines connected to a septic tank system or underground drainage system are preferred in urban slum areas. In isolated cases, low cost latrines like pit privy or bore hole latrine may be adopted.

4.6 ULTIMATE DISPOSAL OF SEWAGE

4.6.1 Disposal into a River

When organic wastes are added to a river, suspended matter either tends to settle or is carried away downstream depending upon the hydrography of stream conditions. Aerobic bacteria utilise the pollutant material, in the presence of sunlight and oxygen available in surface water and decompose organic matter. BOD of wastes is satisfied in course of time and flow-distance in a river. River waters absorb oxygen from the atmosphere continuously. Due to dilution, sewage strength is reduced and due to natural reaeration, river waters are purified. Water may not be fit for drinking purposes in some reaches of river where the pollution is significant. Waste treatment reduces the pollution load on rivers and improves self purification property of natural waters. Pretreatment of wastes removing floating and suspended solids to some extent, prevents the deposition and foam formation near shores.

If the dilution of 300 times or more is available in the river, preliminary treatment of the sewage is sufficient. If the dilution is less than 150 times the quantity of sewage, biological treatment is essential. Sewage effluents discharged into surface water sources should not have more than 20–30 mg/L of BOD and 30–50 mg/L of suspended solids.

Marine waters have less dissolved oxygen, low settling capacity, high density, more tidal currents and wind action, and cause foam formation. But sewage may be disposed 1 or 2 km deep into the sea.

Lake waters are less turbulent and have less mixing capacity, less reaeration capacity, more deposition of solids, and presence of nutrients (N and P) which cause eutrophication (i.e., growth of algal and plant life).

4.6.2 Factors of Self Purification

(a) *Dilution.* In this process, volume of water is comparatively very high in comparison with quantity of sewage. It means, concentration of impurities are reduced. These waters also maintain more dissolved oxygen to satisfy the BOD of sewage.

(b) *Dispersion and mixing.* Strong currents and turbulence in a stream cause dispersion of solids. The thorough mixing of sewage and water helps in the decomposition of the sewage due to biochemical reaction. The reaeration capacity of water is increased due to more surface area of exposure.

(c) *Sedimentation.* Heavier solids settle at a lower level due to natural gravitational forces. But they may be carried away by the stream owing to the high velocity of the stream.

(d) *Temperature.* Warm and summer temperatures are good for sedimentation and biological activity. Solubility of oxygen depends on the water temperature. Temperature difference between water layers also causes movements.

(e) *Sunlight.* The presence of sunlight is required for algal growth and in the process extra oxygen is produced which is needed for decomposition of organic matter. Pathogens are also killed to certain extent.

(f) *Oxidation.* Aerobic decomposition converts organic matter into stable compounds. When the rate of supply of oxygen cannot satisfy the BOD of sewage, the stream conditions may turn anaerobic.

(g) *Reduction.* Hydrolysis of organic matter takes place under anaerobic conditions, when the air supply is inadequate. Methane, hydrogen sulphide, carbon dioxide are released.

(h) *Bottom deposits.* There will be anaerobic conditions at the bottom portions of a river. If disturbed, they may create nuisance and foul conditions.

Self purification is due to the presence of saprophytic and nitrifying bacteria. Organic matter is synthesised with stable end products. Sufficient oxygen (dilution) must be available to save fish and aquatic life.

Due to the presence of nitrogen and phosphates, eutrophication takes place, making waters turbid with less possibility of reaeration. Decomposition of sewage is a natural process, (aerobic or anaerobic) depending on the available conditions. Water absorbs oxygen from the atmosphere. Appearance of crustaceans, rotifers, algae, diatoms, fish, flora and fauna and different species indicate the stage of decomposition and purification level of water.

4.6.3 Oxygen Sag Curve

Oxidation of organic matter requires the presence of dissolved oxygen, sunlight, bacteria, plant life, favourable temperature and other atmospheric conditions. The depletion of oxygen is a first order removal rate. The rate varies with temperature, settling of solids, respiration of plants and animal life, additional runoff, products of digestion, etc. Simultaneously oxygen is absorbed from the atmosphere for respiration process and oxygen is also released due to photosynthesis in the presence of algae and sunlight. Reaeration rate is a function depending upon the rate of water flow, surface area exposed, turbulence, DO saturation level, temperature, photosynthesis and finally the time and distance of river stretch after addition of sewage. The ratio (K_R/K_D) of reoxygenation coefficient and the deoxygenation coefficient is known as self purification capacity of a river. K_R/K_D signifies the pollution level at various stages along the length of the river, at different times. The net balance of oxygen is shown as oxygen sag curve (Fig. 4.36). The curve shows the position of maximum or critical deficit of oxygen in river water, and the point of inflection where the recovery rate of O_2 is maximum. As mixing or initial stages are worst polluted zones, they should be as short as possible.

4.6.4 Zones of Pollution

There are four zones of pollution in river water.

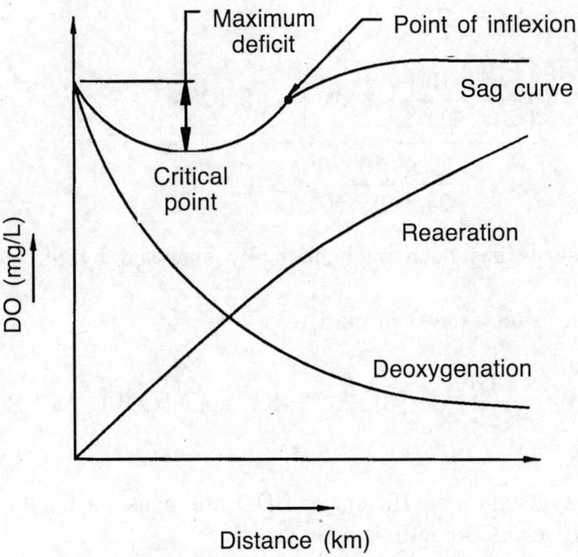

Fig. 4.36 Oxygen sag curve.

1. *Zone of degradation.* In this zone, there is a reduction of oxygen level initially, due to decomposition of organic matter.
2. *Zone of active decomposition.* In this zone, dissolved oxygen is at a minimum (critical) level. Bottom deposits release foul gases and odours. Bacteria and fungi thrive. There is also an increase in ammoniacal nitrogen which makes the fish die.
3. *Zone of recovery.* Reaeration exceeds rate of removal of oxygen rate, and ammonia nitrogen is converted biochemically to nitrates. Rotifers, crustaceans, and other tolerant varieties of biological life appear. The algae flourish on the nutrients gained from the stabilisation of organic matter.
4. *Zone of clear water.* Dissolved oxygen starts to come up to the original level of saturation. Variety of aquatic plants and animals thrive and inorganic compounds (nitrates, phosphates, dissolved salts) increase in river waters.

The characteristics of river also vary with the depth of water—the top layers being aerobic and bottom layers, anaerobic.

Oxygen measure is a yardstick of pollution status of river. Other parameters like turbidity, colour, aquatic life, solids, etc., can also be taken as indicators of pollution.

PROBLEM 14 5 MLD of sewage with 200 mg/L of BOD is discharged into a river having DO of 8 mg/L and a rate of flow of 200 L/s. Find the BOD of the mixture.

Solution

$$\frac{\left(\frac{5 \times 10^6}{24 \times 60 \times 60} \times 200\right) - (8 \times 200)}{\frac{5 \times 10^6}{24 \times 60 \times 60} + 200} = 37 \text{ mg/L}$$

Oxygen sag curve has been mathematically analysed by Streeter-Phelps and expressed as:

DO deficit (after t days) in mg/L =

$$\frac{L.K_D}{K_R - K_D}[10^{-K_D.t} - 10^{-K_R.t}] + D_O(10)^{-K_R.t}$$

where,

K_D = deoxygenation coefficient or BOD rate constant for the waste water
K_R = reoxygenation coefficient for the river
L = ultimate first stage BOD of the mix at the point waste discharge (mg/L)
D_O = initial oxygen deficit of the mix at the mixing point (mg/L)

For streams of normal flow, K_R may be taken as 0.2 and K_D waste water as 0.1. K_R/K_D is known as self purification constant (f). The initial time (t_c) after which the minimum dissolved oxygen occurs, is given by

$$t_c = \frac{1}{K_D(f - 1)} \log\left[\left\{1 - (f - 1)\frac{D_O}{L}\right\} f\right]$$

and the critical or maximum oxygen deficit (D_c) is given by

$$D_c = \frac{L}{f}[10]^{-K_D t}$$

PROBLEM 15 Determine the critical oxygen deficit and its location given that D_O = 2.5 mg/L; K_D = 0.1; K_R = 0.20, L = 50 mg/L and flow velocity = 0.15 m/s.

Solution

$$t_c = \frac{1}{0.1(2 - 1)} \log\left[\left\{1 - (2 - 1)\frac{2.5}{50}\right\} 2\right] = 2.79 \text{ days}$$

$$D_c = \frac{50}{2}[10]^{-0.1 \times 2.79} = 13.15 \text{ mg/L}$$

Distance of critical deficit = 0.15 × 2.79 × 24 × 60 × 60 = 36.16 km.

4.6.5 Disposal on Land

The treated effluents are discharged on land for irrigation purposes if BOD (200 mg/L), pH value (5.5 to 9), TDS (1500 mg/L), chlorides (500 mg/L), and sulphates (1000 mg/L) are within tolerance limits. While successfully disposing sewage, sewage farming encourages growing of crops (which are not eaten raw and not in contact with sewage directly), such as grass, cotton, sugarcane, papaya. Various techniques are used such as broad irrigation, subsurface irrigation, spray irrigation. A ridge-and-furrow type of irrigation is shown in Fig. 4.37.

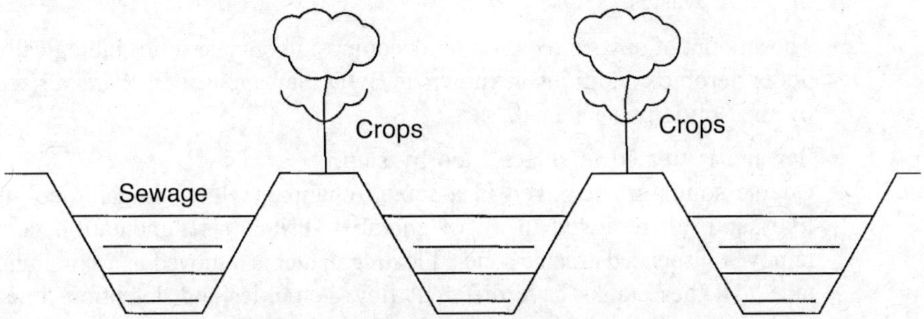

Fig. 4.37 Ridge-and-furrow method of irrigation.

Water application rate depends upon crop requirements, climate, soil characteristics, etc. Normally 100–200 m^3/day/hectare may be applied, whereas BOD load can be 50–60 kg/day/hectare.

When sewage gets deposited continuously on land, soil gets clogged with sewage solids and the land is not in a position to receive further load. This sickness of soil is prevented by primary treatment of sewage, resting of land, rotation of crops, ploughing the land, and improving soil drainage.

4.6.6 Land vs River Disposal

The following points may be noted on the ultimate disposal of sewage.

1. If rivers are nearby and sufficient quantities of dilution are available, sewage is disposed on water bodies.
2. If large areas of porous (sandy) land is available, near urban areas, at a low price, land disposal method is preferred.
3. Other factors to be studied are ground water table, soil permeability, climate, pre-treatment of sewage, sources of drinking water, the river flow in summer months, and the sanctity of rivers etc.

SUMMARY

- Sewage consists of faecal matter and other waste waters from domestic use. Sewage contaminates water sources. Organic matter in sewage decomposes, depleting oxygen from water sources. Hence sewage needs adequate treatment before disposal onto land surfaces or into the water bodies.
- The water carriage system (i.e. underground drainage system) is essential for any residential area, for the safe collection, conveyance and disposal of liquid wastes.
- The amount of oxygen required for decomposing organic solids biologically under aerobic conditions is known as BOD. Sewage is also characterised by the solids present in it.
- Sewage is treated in stages, step by step.
  Coarse solids are removed in a screen chamber while inorganic solids like sand are removed in a grit chamber. Primary sedimentation tank removes suspended organic solids. Floating matter is removed in a skimming tank. All these tanks have different flow velocities and detention times such that only specific solids are retained in that unit.
- Biological action is categorised under aerobic or anaerobic conditions. Aerobic bacteria decompose (oxidise) sewage into CO_2 and water in the presence of oxygen. Trickling filter, activated sludge treatment plant, oxidation ponds, are aerobic biological units.
- Anaerobic decomposition produces gases such as methane, hydrogen sulphide, ammonia. Sludge digestion is an anaerobic process. After digestion, dewatered sludge is disposed.
- Trickling filters are coarse granite beds where sewage comes into contact with filtering media. Organic matter is oxidised in the presence of oxygen by aerobic bacteria attached to the filtering media.
- In activated sludge treatment process, sewage is abundantly aerated, mixed with prior activated (aerated) sludge, seeded with microorganisms. Biological oxidation is comparatively superior to that in a trickling filter unit. A number of modifications in the aeration systems improve the process.
- Oxidation ponds are low-cost, holding basins for the treatment of domestic and industrial waste waters (using algae and bacteria).
- After treatment, sewage is ultimately disposed on land (to be used as manure) or in river water (utilising the dilution capacity of streams).

REVIEW QUESTIONS

1. List the advantages of a water carriage system.
2. Write a procedure for estimating sewage and storm water flows.
3. How are sewers designed?
4. Sketch the following:
   (a) Drop manhole
   (b) Overflow weir
   (c) Catch pit
   (d) Ejector
5. Explain the terms BOD and decomposition of sewage.
6. Explain the self purification property of rivers and the various stages of an oxygen sag curve.
7. With a flow diagram, explain the functions of each unit in primary treatment of sewage.
8. Explain the working principles of
   (a) Oxidation ponds
   (b) Trickling filters
   (c) Septic tank
9. Describe the activated sludge treatment process (ASTP) and state the merits of ASTP over trickling filters.
10. Explain the process of sludge digestion and state the factors affecting the same.
11. Describe briefly the various methods adopted in final disposal of waste water sludges.
12. Write short notes on
    (a) House drainage
    (b) Oxidation ditch
    (c) Recirculation
    (d) Land disposal
    (e) Microorganisms and their characteristics.

OBJECTIVE TYPE QUESTIONS

A. ***Each of the following questions comprises four alternative answers. Select the most appropriate answer for each.***

1. Water seal is provided in a
   (a) Wash basin (b) Urinal
   (c) Closet (d) Flushing tank

2. The sewer preferred to develop self cleaning velocity during minimum flow is
   (a) Horse shoe type (b) Circular type
   (c) Egg shaped (d) Box type
3. Fertilising constituents of sewage are
   (a) Nitrogen and iron (b) Iron and phosphorus
   (c) Phosphorus and nitrogen (d) Chlorides and sulphates
4. If the sewage is fully oxidised, the nitrogen is in the form of
   (a) Nitrites (b) Ammonia
   (c) Nitrates (d) Nitramines
5. Commonly adopted size for house drains and sewers is
   (a) 75 mm (b) 50 mm
   (c) 100 mm (d) 125 mm
6. Because of sludge bulking,
   (a) gas volume increases in digestion unit
   (b) sludge settles quickly in clarifiers
   (c) sludge doesn't settle easily
   (d) all the above
7. In an activated sludge treatment process,
   (a) Settling takes place in an aeration unit
   (b) Air is introduced for a period of 4 to 8 hours, daily
   (c) Agitation of sewage is done by air
   (d) New microorganisms are added daily.
8. Which treatment unit is designed on the basis of 0.036 m^3/head (volume)
   (a) Sedimentation tank (b) Trickling filter
   (c) Digestion unit (d) Septic tank
9. The following statement is wrong, with reference to sludge digestion.
   (a) Rate of digestion is more at higher temperatures
   (b) pH value of 5 is not desirable for methane formation
   (c) Digestion period is about 15 days with an optimum temperature of about 30°C
   (d) Mixing the contents improves decomposition.
10. In a trickling filter, if the recirculation factor (F) is unity, then the ratio, volume of sewage recirculated to the volume of raw sewage is
    (a) 1 (b) 0.5
    (c) 0 (d) 1.5
11. A surface loading of 2000 L/m^2/day is adopted in
    (a) Sedimentation tank (b) Digestion unit
    (c) Trickling filter (d) Activated sludge process
12. Settling velocity is proportional to square of the diameter of the particle (according to Stoke's Law), when
    (a) Diameter of particles is between 0.1 and 1.0 mm
    (b) Diameter of particles is more than 1.0 mm

(c) Diameter of particles is less than 0.1 mm
(d) For all diameters

13. Find the statement not related to sewage farming
(a) Porous soil (b) Low ground water table
(c) Green vegetables (d) Land sickness

14. BOD curve is similar to
(a) Reoxygenation curve in river waters
(b) Oxygen sag curve
(c) Deoxygenation curve in polluted rivers
(d) Logistic curve

15. Absence of fish along a river indicates
(a) Zone of degradation (b) Zone of active decomposition
(c) Zone of recovery (d) All zones of pollution

16. If an industry contributes 320 kg/day of BOD, its population equivalent is
(a) 8000 (b) 3200
(c) 4000 (d) 40,000

B. *Fill in the blanks with suitable words.*

1. Standard BOD curve is drawn with reference to time and ________.
2. The unit which removes grease from sewage is ________.
3. First stage of BOD is known as ________.
4. Settleable solids are measured by ________.
5. The average BOD_5 of domestic sewage is ________ kg/day/head.
6. 30 cm diameter sewer is laid with a slope ________ so as to maintain a self cleansing velocity (0.8 m/s)
7. 1 L sewage gives 2.7 cm^3 of settled sludge in 30 minutes. If the dry weight of sludge is 3 g, its sludge volume index (SVI) is ________.
8. When 30 cm/s flow velocity is constantly maintained in a settling unit (with about 1 minute detention time), ________ is removed.
9. The rate of application of sewage on to the land is about ________ m^3/hectare/day.
10. The basic photosynthetic (chemical) equation is ________.
11. In a grease trap, outlet level is located at ________.
12. Vitrified clay pipes are widely used in ________.
13. No deposition is allowed in sewers because ________.

CHAPTER 5

Waste Water from Industries

5.1 POLLUTANTS

Solid, liquid and gaseous effluents from various industries are generally disposed into the nearby environment. Almost all industries use water for some process or the other and finally discharge the waste waters. Pollutants generated by industries include (a) acids and alkalies (b) suspended solids (c) toxic organic chemicals (d) toxic inorganic chemicals (e) non-toxic organic chemicals (f) non-toxic inorganic chemicals (g) materials that produce colour, turbidity, odour and taste (h) oil and grease (i) radioactive material (j) high temperature wastes and (k) non-biodegradable substances. Pollutants from industries can also be classified as (a) pH changing, (b) oxygen consuming, (c) non-biodegradable substances, and (d) easily biodegradable substances. These pollutants should undergo a purification process within the factory boundary limits and suitable precautionary measures are to be adopted for their disposal. Otherwise, the total surrounding human environment gets polluted beyond rectification.

Biochemical oxygen demand (BOD) caused by the liquid effluents may increase even beyond 30,000 mg/L whereas the domestic sewage exerts approximately 200 mg/L of BOD. pH value of industrial effluents may vary from 3 to 11, (i.e. extreme conditions of acidity or alkalinity) whereas pH of domestic wastes is near neutral value. Similarly, total solid content of industrial wastes which also contains toxic chemicals, is very high.

5.1.1 Harmful Effects

Some undesirable characteristics of waste water are:

1. Soluble organics cause depletion of oxygen in river water.
2. Suspended solids impair normal aquatic life, form benthal deposits, and may get decomposed.
3. Trace organics and phenols cause tastes and odours.

4. Heavy metals, cyanides and toxic substances are harmful to living groups.
5. Colour and turbidity create unaesthetic conditions in water bodies. As a result, oxygen availability and photosynthesis are also affected.
6. Nitrogen and phosphorus stimulate undesirable algal growth and enhance eutrophication.
7. Refractories resist biodegradation.
8. Oil and floating matter are unsightly and retard reaeration.
9. Alkyl benzene sulphonate (detergents) leads to foam formation in river water.
10. Hot water discharges from factories act on solubility and affect bacteriological action.
11. Acids and alkalies affect fish and aquatic life.
12. Inorganic materials may give rise to various problems such as hardness, and corrosion.

5.1.2 Waste Characteristics

Some of the harmful chemicals from the industries are arsenic, cadmium, chromium, lead, mercury, cyanide, nitrogen, iron, zinc, and metallic sulphates, chlorides, nitrates, fluorides, phosphates and sulphides. Table 5.1 gives a fair idea of the characteristics of the untreated effluents from various industries.

1. Ratio of BOD to COD is 0.5–0.6 in dairy, distillery and textile wastes. If the ratio is greater than 0.5 or 0.6, wastes are biodegradable. For pharmaceutical wastes, this ratio is nearly 1.5 and about 0.15–0.20 for petrochemical and steel units. Sometimes, wastes are treated biologically after acclimatisation.
2. Industry specific effluent standards are prescribed at the national level, designated as Minimum National Standards (MINAS) so as to envisage uniform standards for treatment and disposal at various stages. The standards permissible for waste disposal into rivers, on land disposal, and for disposal into public sewers are also prescribed separately.

5.2 MIXING OF INDUSTRIAL AND DOMESTIC WASTES

Domestic sewage and industrial wastes may be treated separately or jointly, before the final disposal into large bodies of water or on waste lands. Decision has to be taken only after studying (a) waste quality and quantity, (b) whether treatment required should be separate or combined and (c) receiving end conditions such as quantity of dilution available or availability of waste lands. The following points would indicate if mixing of the two types of wastes is to be carried out before the final disposal.

Table 5.1 Characteristics of Industrial Effluents

Sl. No.	Characteristics → Source ↓	Waste Volume	BOD. (mg/L)	pH	Solids (mg/L)	N (mg/L)	Chlorides (mg/L)	Others
1	Dairy	5–10 times of milk production	1000	8	1000	—	—	Oil and grease, odours, readily putrescible, biodegradable
2	Distilleries	30 L/litre of Alcohol produced (or 150 m^3/day)	Upto 40000	4	50000	1000	5000	Dark brown, dissolved solids, iron, sulphates, phosphates
3	Fertiliser	10000 m^3/day or 10 m^3/tonne	500	8.5	2000	10000	4000	As, F (10 mg/L), NH_3, phosphates (1000 mg/L), SO_4 (5000 mg/L)
4	Paper	300 m^3/tonne of pulp produced	400	8–9	1500	–	300	Brown, fibrous matter, lignin, caustic soda
5	Petrochemicals	20 m^3/tonne of crude refined	500 to 3000	Varies	500–5000	200	2000	Oils, phenols, sulphides, cyanides, hydrocarbons
6	Pharmaceuticals	100 m^3/day	3000 to 8000	4 or 9	1500–10000	500	—	Toxic organics, strong odours, sulphides, phosphates
7	Steel	2 m^3/tonne of iron produced	2000	5–8	2000	300	—	Phenol, cyanides, phosphates, SO_4 (500 mg/L)
8	Tanneries	100 m^3/tonne of raw material	2500 to 5000	8–9	10000	—	5000	Strong odour and colour, high salt/organic content, sulphates, chromium

(Contd.)

9	Textiles	200 m^3/tonne of cloth	1000	9–10	8000	30	—	Colour, sulphates, Ca, Mg, Na, Zn
10	Domestic sewage	200 L/day/ capita (approx.)	300	6–8	1000	—	—	—
11	Max. permissible limits for discharging into surface waters	—	30	5.5–9	S.S100	100	—	Phenol(1.0), F(2), oil/grease(10), Cr(0.1), Hg(0.01), CN(0.2), As(0.2), Cu(3), Pb(0.1), Ni(3), Se(0.05), sulphides(2)

Note: Data is subjected to variations from plant to plant.

1. If the waste mixture (domestic and industrial) (i) is homogeneous in composition, and is uniform in flow rate, (ii) has less suspended solids and BOD, (iii) is in the near neutral zone of pH, and free from toxic chemical content, then industrial waste may be mixed with domestic sewage for combined treatment before ultimate disposal into stream.
2. If some constituents are removed or partial treatment is given to wastes from industries (such as removal of toxic chromium), then it can be mixed with domestic sewage for further treatment.
3. Industrial wastes may be treated completely and discharged into domestic sewers, to serve as an additional protection before final disposal.
4. Raw wastes should not be directly thrown into rivers unless pH, solids, toxic substances, BOD. and other pollution characteristics are properly analysed. Stream conditions such as water quality, flow rate and quantity should also be examined to assess self purification capacity of the river water.
5. Partially treated industrial wastes may be disposed into rivers depending upon the dilution capacity of the water and the pollution load on it.
6. Even if the completely treated industrial wastes (after satisfying the permissible limits for discharging the waste waters into rivers) are disposed, the river water should be purified before being used for drinking purposes.

5.3 PRE-TREATMENT OF INDUSTRIAL WASTES

The following steps are essential for the pre-treatment of industrial wastes.

5.3.1 Reduction of Waste Strength and Volume

Strong and weak wastes may be separately collected, treated and disposed as mixing the two types of wastes create large volumes and treatment becomes difficult. If dilution is advantageous, they may be combined.

Classification and grouping of different categories of wastes would help the engineer to take a suitable decision on the treatment and disposal methods.

1. Disposal of cooling waters (used in steel plants), floor washings, and sullage waters is easy. Separation of liquid kiering wastes from textile mills and separation of chromium from metal finishing units are also not difficult.
2. Number of useful substances like grease, caustic soda, silver, potash, chromium may be recovered easily for economical gains and to reduce pollution load.
3. Waste water may be conserved by reusing or recycling within the industry for example—white liquors in paper industry, cooling waters in steel plants.

4. Process and equipment modifications, may be tried to reduce the strength of waste. Use of phosphoric acid instead of sulphuric acid in pickling units, use of carboxymethyl cellulose in place of starch in textile mills, reduction of cyanide in metal plating industries, are quite useful steps in the production stages to reduce pollution.

5.3.2 Equalisation and Neutralisation

Treatment units are always designed for a specific influent characteristics and composition of wastes. Fluctuations or variations in the input quality affect the purification process. So, consistency in the quality should be maintained by storing the liquid wastes for a certain period of time and mixing suitably the various components in a tank. From this unit, (i.e. an equalisation tank or pond) uniform nature of the influent (with regard to BOD or solids) is fed into the treatment units.

Most of the biological treatment plants function effectively within a range of 6–9 pH. So industrial wastes of acidic or alkaline nature are neutralised in a tank or pond. Sometimes waste waters from nearby units are mixed to obtain neutrality. For example, acid wastes may be passed through beds of limestone. Waste fuel gas, as carbonic acid from boiler units, may be allowed to react with caustic wastes to obtain a pH value of seven.

When industrial wastes and domestic sewage are mixed for combined treatment, separate storage tanks help to properly proportion and regulate the quality and quantity of wastes into a purification plant. Industrial wastes may be partially treated and disposed into domestic sewers for further treatment.

5.4 VARIOUS INDUSTRIES

Process operations, pollutant characteristics and purification procedures for various industries are presented briefly.

5.4.1 Dairy Plant

A dairy plant consists of milk receiving unit and Pasteurisation plant, for manufacture of butter and dairy products. Effluents come from washing operations and are likely to contain residual milk and used chemicals (detergents). Milk and milk products, readily putrescible fats, proteins and lactose, strong organic and biodegradable substances come from dairy plants. High rate trickling filters and activated sludge plants are efficiently used as aeration can reduce BOD to a good extent. Oxidation ponds are also designed to maintain aerobic conditions for treatment, at a much lower cost.

5.4.2 Cane Sugar and Distilleries

Lime is added to crushed sugarcane. The Juice of sugarcane is filtered, evaporated and crystallised to obtain sugar. In distilleries, mother liquor (molasses) is acidified with sulphuric acid, supplemented with nitrogen and phosphorus, fermented with yeast and finally distilled for obtaining ethanol. About 3000 L/day of waste water is generated during the production of one tonne of sugar. Distillery wastes from molasses, containing high dissolved salts and ash content, are acidic and have high BOD. Total waste water volume is 50 L. per litre of alcohol produced, and originates mostly from stillage and fermenters.

Bagasse from sugar plant is useful as a fuel, and for preparation of fibre boards. Potash can be recovered from spent waste. Distillery waste is treated by lagoons followed by aerobic oxidation. Anaerobic digesters are especially useful as most of the organics are converted into CO_2, CH_4 and other products in anaerobic digestion. Biogas and fertiliser (or cattle feed) can thus be obtained from this process.

5.4.3 Fertiliser Industry

Fertiliser units mainly produce nitrogenous or phosphate compounds or a combination of these. The presence of phosphates and nitrogenous compounds in the waste waters from fertiliser plants depends upon the nature of fertilisers being produced in the plant. Fluorides, oil and grease, arsenic etc. also join the wastes.

500 litres of liquid effluent is approximately generated in the production of one tonne of urea. In nitrogenous and phosphatic fertiliser industries, sulphuric, nitric and phosphoric acids, CO_2, ammonia, rock phosphates, coal, naphtha, etc. are used as raw materials. Carbon is removed and used again. Oil is removed by oil separators followed by coke absorption. Lime addition and clariflocculation neutralises and precipitates impurities like fluorides, and phosphates. Air stripping helps in the recovery of acids. Ammonia and ethanol are stripped off in a distillation tower by means of steam. Air stripping at high pH, dissociates ammonia, and further converts ammonia into nitrates. It is anaerobically converted into nitrogen and finally sludge is removed in lagoons or settling tanks.

5.4.4 Paper Industry

Cellulose containing wood material is cooked in caustic soda, sodium sulphide mixture, under controlled temperature and pressure. Spent liquor (black liquor) is used for recovery of caustic soda, while pulp is washed and bleached with chlorine. In the process, sulphate is reduced to sulphide and is converted to sodium hydroxide with lime treatment. The white liquor thus obtained is sent to digestors. Pulp is used in paper manufacturing by adding dyes, alum and talc to fibres. Large volume of water is required to dilute pulp. Recovery of chemicals and fibres reduces the pollution load. Lignin is recovered from black liquor after precipitating with lime addition and alum coagulation treatment. Sedimentation and flotation is a reliable treatment. Lagooning is an economical treatment for the removal of 80% BOD. Activated sludge treatment is also a satisfactory treatment process for paper and pulp mill wastes. Nutrients (N, P) may be added (if required) to help the treatment process.

5.4.5 Refineries and Petrochemicals

Processing and cooling operations generate a waste volume of 20 m^3 per 1 tonne of crude, in oil refinery unit. Major pollutants are oil (500 mg/L) and chemicals such as acids, alkalies, sulphides (50 mg/L), phenols (50 mg/L), BOD, (500 mg/L).

Depending upon the chemical process and manufactured item, pH, BOD, solids, sulphides, phenols, etc. are to be estimated. API (American Petroleum Institute) gravity separators, and sedimentation cum air flotation tanks, remove oil substances. For emulsified oils, coagulation method is used while sulphides are removed by aeration. Biological (aerobic) treatment is also very effective. Oxidation ponds remove BOD, sulphides, phenols and oils to a satisfactory level.

Air flotation and clarifloculators separate free as well as emulsified oil. Neutralisation, activated sludge units and oxidation ponds are suggested units of treatment. Cyanides, fluorides and sulphides are segregated and treated. Sulphides are precipitated by the addition of ferric chloride. Fluorides are precipitated by the addition of lime or absorbed using activated alumina or activated alum treated carbon. Cyanides are removed by alkaline chlorination and later treated in aerobic biological units.

5.4.6 Pharmaceuticals

Basic drugs and various formulations are manufactured in pharmaceutical units. Waste water quantity and characteristics depend on the nature of products, processes, raw materials and purification process of finished products. Antibiotic plant yields a waste water with pH value of 4.5 whereas sulpha drug unit results in waste water with pH of 9. BOD varies from 1000 to 10,000 mg/L and total solids exceeding from 10,000 to 50,000 mg/L. Wastes mostly come from filtration and distillation units. Neutralisation, extended aeration, chemical oxidation, filtration, and oxidation ponds are employed for treating pharmaceutical wastes. Chemical treatment with ferrous sulphate or alum is also useful. Although anaerobic lagoons are widely used, aeration units followed by clarifiers are also adopted in some cases. When total dissolved solids content is high, multiple effect evaporators are used. Hazardous wastes are generally incinerated.

5.4.7 Steel Plants

Steel plants have material handling unit, coke ovens, blast furnaces, other furnaces, casting, rolling, pickling, heat treatment and finishing operations. Water is used in different processes in cooling and washing systems. Wastes from coal washeries and blast furnace units are voluminous. A number of chemicals such as lead, zinc, chromates, phosphates, chlorides, sulphides, acids, alkalies, iron, silica, magnesium, phenols and cyanides may appear in the liquid effluent.

Froth flotation process removes fine coal particles. Clarifiers with or without coagulation units are used. Chemical oxidation and biological treatment are adopted.

5.4.8 Tanneries

Process operations include soaking raw hides in water, treating with lime and sodium sulphide for removal of hair and flesh, treating hides in vegetable tannin solution of chromium sulphate liquor solution for tanning, washing, neutralising and finally treating with emulsion of sulphonated oils and dyeing. Processes used are classified as pre-tanning, tanning and finishing operations.

Effluents contain higher amounts of suspended and dissolved solids resulting in strong colour and odour. They are highly alkaline. After the recovery of chromium, wastes are treated by various units of operations like screens, sedimentation tanks, conventional lagoons (aerobic, anaerobic, or in combination). Activated sludge treatment process, trickling filter units or oxidation ditches are very effective in reducing of BOD. Chemical precipitation is not desirable due to problems involved in handling sludge.

5.4.9 Textile Units

Textile mills generally produce cotton, woollen or synthetic fibres. Desizing, scouring, bleaching, mercerising, dyeing and printing operations of cotton mills generate liquid wastes to the tune of 3000 m^3/day in a medium sized plant. The pH value of a cotton textile waste is 8–10, whereas it is 3–5 for wastes from a synthetic fibre unit. Organic, as well as inorganic substances, starch, cellulose, sodium hydroxide, detergents, hypochlorites, dyes, pigments, sodium gums, waxes, sulphides and sulphates are present in waste waters. Primary treatment consists of segregation, recovery, recycling of water, screening, neutralisation with lime, equalisation to maintaining flow balance and uniform characteristics, chemical coagulation with lime and iron salts and clarification for colour removal. Finally, biological units viz. trickling filter and activated sludge treatment plant are essential for the final treatment of textile wastes.

Flow diagrams of treatment units for selected industries are shown in Figs. 5.1–5.7.

5.4.10 Other Industrial Units

1. Liquid wastes from electroplating and metal finishing units contain Ni, Cr, Zn, Pb, Ag, Cd, Hg, CN, etc. depending upon the manufacturing unit. Treatment option is to recover metallic ions, reuse water and remove toxic chemicals.
2. Water pollution is due to wash-off from discharge of mine water, acid mine discharge (due to sulphide minerals present in coal and metallic ore bodies) and coal washing operations in mining units. Neutralisation with lime, soil conservation and biological reclamation of land are desired.
3. A number of industries manufacturing pesticides and metallurgical units also generate liquid effluents which ultimately pollute soil and water bodies.

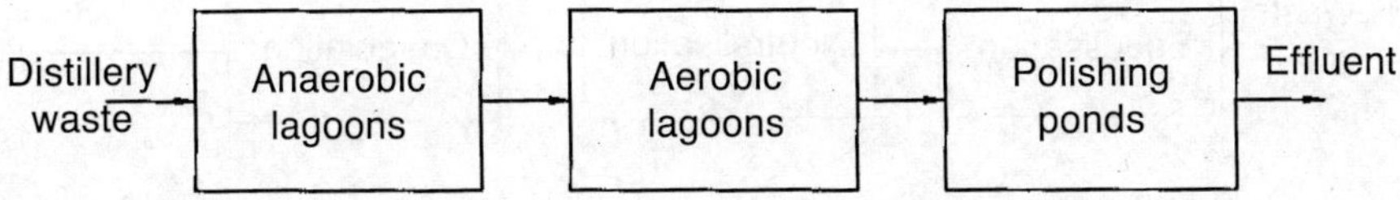

Fig. 5.1 Distillery water treatment.

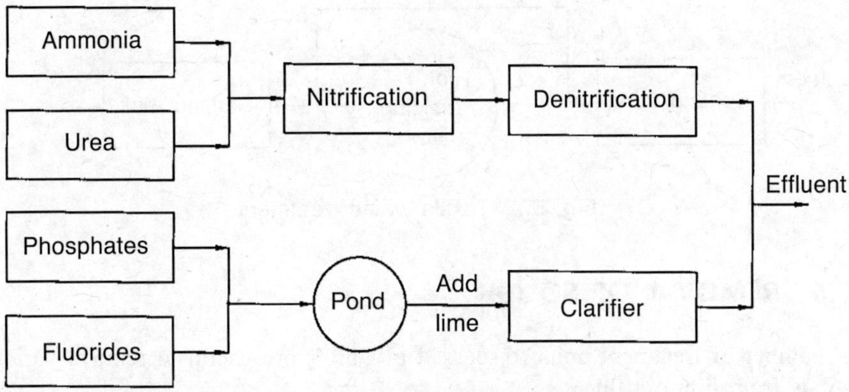

Fig. 5.2 Fertiliser plant waste treatment.

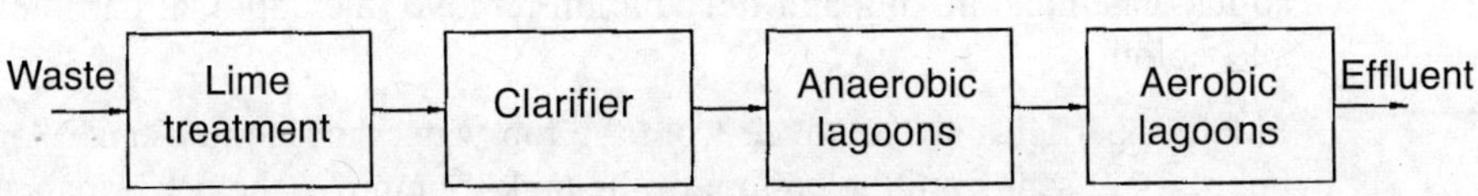

Fig. 5.3 Paper mill waste treatment.

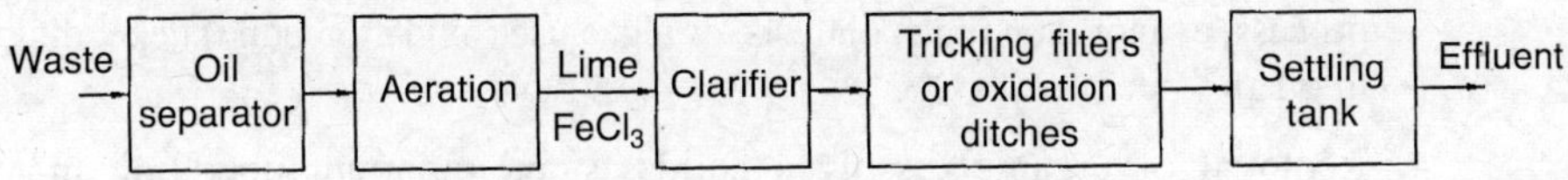

Fig. 5.4 Petrochemical waste treatment.

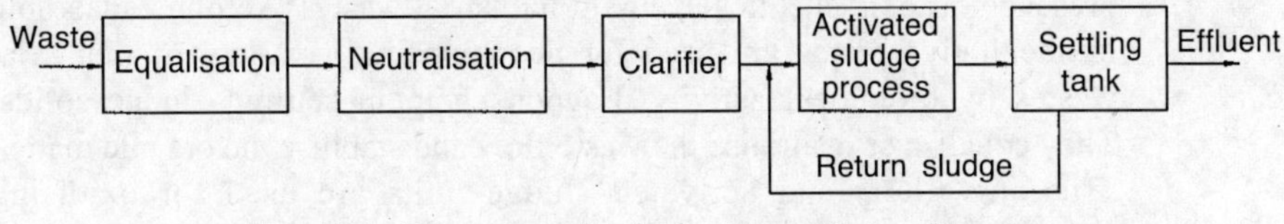

Fig. 5.5 Pharmaceutical plants waste treatment.

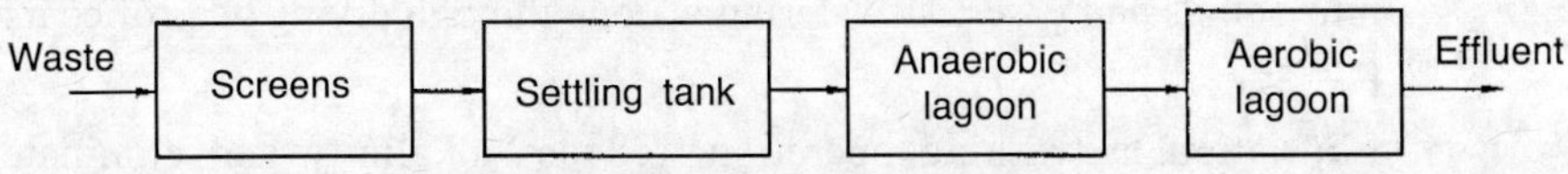

Fig. 5.6 Tannery waste treatment.

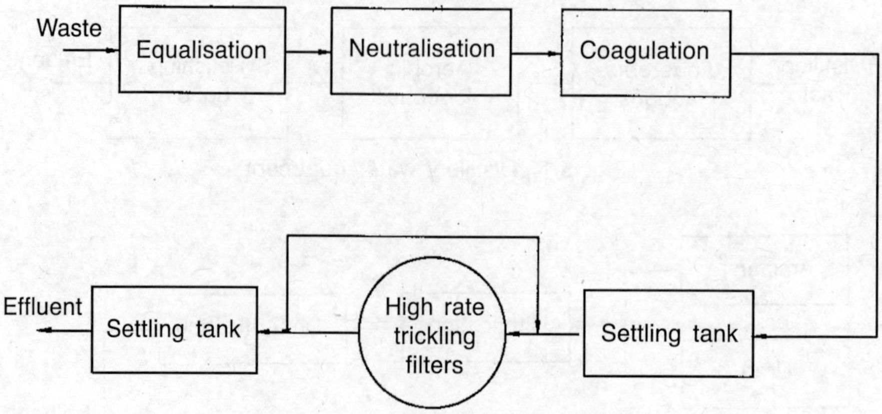

Fig. 5.7 Textile waste treatment.

5.5 REMOVAL OF SOLIDS

A number of treatment units (discussed in Ch. 4) are useful to remove various solids from liquid effluents.

1. Sedimentation tank removes a good percentage of settleable suspended solids. Chemical mixing and flocculation remove fine colloidal particles and colour.
2. Floating particles are removed from the top water surface in a tank. Air bubbles are sent from the bottom of the tank to improve the performance of flotation tank.
3. Inorganic dissolved salts are removed by an ion exchange (demineralisation) process, evaporation units, dialysis, or algae use (oxidation pond) depending upon the need.
4. Removal of organic dissolved solids is an important operation in a waste treatment process utilising the biological activity. But micro-organisms are sensitive to temperature, pH value, oxygen levels, presence of toxic chemicals, etc. in industrial wastes. Aerobic, anaerobic or facultative ponds are used for decomposing a variety of industrial wastes by biological activity. Lagoons help in settling sludge solids. They equalise or neutralise the waste flow and stabilise the organic matter. Trickling filters and activated sludge units are used for oxidising organic matter and removing BOD. Anaerobic sludge digestion helps in reducing the solid volume, imparting drainability and converting solids into liquids and gases. Dewatering is done through drying beds of coarse gravel.
5. Elutriation, a washing operation, reduces alkalinity and eliminates odours generated from waste solids.

5.6 SPECIAL PROCESSES

In addition to the conventional waste water treatment methods, industrial wastes need processes to remove chromium, phenol, mercury, nitrogen when they are present in excess quantities.

Chemical treatment is not commonly used in domestic sewage treatment. In treating various types of industrial waste waters, pre-treatment with chemicals becomes necessary before biological treatment using mixed culture micro-chemical coagulation (for fine suspended and colloidal solids which cannot be removed by sedimentation or flotation), for the removal of phosphorus, toxic and coloured substances, and soluble organics. Heavy metals like Ni, Zn are chemically precipitated first as hydronides and then removed. Lead is precipitated as carbonate or phosphate. Fluoride is precipitated as Ca or Al fluoride. Phosphates are removed as tricalcium phosphate using lime. Cyanide (CN) is removed by treatment with alkali and chlorine.

Manganese sulphate is used as a catalyst for removing H_2S by aeration from tannery and oil refineries waste waters. Adsorption is used for removing colour, phenol, etc. Activated carbon is most commonly used to remove pollutants from industries compared to domestic sewage. Various treatment methods like physical treatment (using screens, grit chambers—both aerated and simple, equalisation and proportioning tanks, grease traps, gravity separators, dissolved air flotation, sedimentation tanks), chemical treatment (pH adjustment, chemical precipitation, chemical coagulation, chemical oxidation and/or reduction), physico-chemical (adsorption) and aerobic and/or anaerobic biological treatment methods (such as tricking filter and activated sludge treatment plant and its modifications, anaerobic attached growth, anaerobic suspended growth and hybrid reactors) are available for treating industrial wastes.

5.6.1 Removal of Chromium

Hexavalent chromium is toxic (0.1 mg/L threshold limit) and is present in tanning, electroplating, fertilizers and other industries manufacturing some organic chemicals. The procedure of removing chromium comprises the following steps.

1. Cr_6 is reduced to Cr_3 by the addition of sulphuric acid (to reduce pH to 2–3) and $FeSO_4$. Later on, it can be precipitated in settling tanks after neutralising with NaOH or $Ca(OH)_2$. This is an economical method.
2. Cationic resins can be used for the recovery of Cr in the form of sodium chromate or chromic acid. Regeneration of the resins is practised with NaCl or H_2SO_4 solutions as the case may be.
3. Lime coagulation (precipitation and settling) and adsorption with activated carbon is also feasible.
4. Sometimes reverse osmosis process with suitable semi-permeable membrane is also adopted.

5.6.2 Removal of Phenol

Phenols are present in the waste waters released by chemical, petroleum, pharmaceutical, plastic, metallurgical, printing, and textile units. The steps followed in the removal of phenol are as follows.

1. Oxidising agents such as hydrogen peroxide, potassium permanganate, and sulphur dioxide, help in reducing phenolic compounds. Aeration and biological action are normal methods of treatment.
2. Steam stripping, adsorption, ion exchange, and solvent extraction are the other methods used.

5.6.3 Removal of Mercury

Chloralkali plants, mining industry, paper and pulp mills and pesticides units are the main sources of mercury. Methyl mercury is more toxic and will magnify through food chain.

1. By adding H_2S or Na_2S, mercury can be precipitated as sulphide which can be removed by settling process. $FeCl_3$ may be added, if necessary.
2. Activated carbon, clays, silica gels are used for adsorption.
3. Mercury salts in solution may be reduced by the addition of active metals like iron.
4. Natural or synthetic resins may be used in the process of ion exchange for the removal of mercury.

5.6.4 Removal of Nitrogen

Nitrogenous compounds (ammonia) enter water bodies (through fertiliser and other industries) causing eutrophication.

1. Low cost treatment to reduce ammonia is to adjust pH and utilise bacteriological and algal symbiosis in a stabilisation pond.
2. Simple and economical treatment is to adjust pH to 10–11 by lime addition and later adopt aeration with large quantities of air (air stripping).
3. Nitrification under aerobic conditions at pH 8–9 and denitrification under anaerobic conditions with suitable microorganisms (nitrosomonas and nitrobacter) are suitable procedures in conventional treatment units.
4. Ion exchange or reverse osmosis process is used selectively for the removal of nitrates or ammonia.

5.6.5 Removal of Dissolved Salts

Removal of inorganic dissolved salts is done in special units which are described below.

(i) *Ion Exchange.* Certain compounds like resins, possess a property to exchange one of their ions from the surrounding medium due to ionisation potential and electrolytic action.

(ii) *Dialysis.* It is the process of separation of solutes by means of their unequal diffusion through semipermeable membranes (due to concentration gradient). Due to electrical potential difference, ions get extracted in the solution during electrodialysis.

(iii) *Reverse Osmosis.* Solvent flows from strong concentrated solution to the dilute salt concentration due to the application of high pressure (higher than osmotic pressure).

(iv) *Algae Use.* In oxidation ponds, algal growth utilises minerals such as N, P, K, Ca, Mg for their food and activity and decompose the compounds to a stable level.

(v) *Multiple Effect Evaporation.* This helps in concentrating mineral salts.

5.7 CONCEPTS OF COMMON EFFLUENT TREATMENT PLANTS (CETP)

Liquid effluents from a group of pharmaceutical or chemical industries (located at one place in an industrial estate) can be collected and treated in combination in one treatment plant. This common effluent treatment plant can be managed effectively and efficiently by a separate team of experts in the field. Industries contribute finances to the functioning of CETP on the basis of pollution load (waste quantity and characteristics). After treatment, waste water may be disposed into a municipal sewer or into water bodies.

In case sewers are not available, liquid effluents may be transported from the industries to CETP, by means of tankers. It is always better if industries have primary units of treatment or even some specific units of treatment within their premises and send wastes to CETP afterwards only.

Wastes should have similar characteristics and be homogenous in nature. Addition of domestic sewage from the neighbouring residential colonies (if any) enhances biological activity in the mixture of industrial wastes and sewage besides dilution. It is possible that COD and TDS may vary from industry to industry. An equalisation tank is essentially the first unit in any CETP. Air flotation process is useful to collect floating matter and supply oxygen to the mixture. Biological treatment may include anaerobic units or aerobic units or a combination of both. Finally, sludge is removed for further treatment and disposal, whereas liquid effluents are sent to oxidation ponds for final disposal. A typical flow diagram is shown in Fig. 5.8.

The following questions need to be answered to assess the feasibility of setting up a CETP.

1. Is it feasible to have one or same line of treatment when the waste characteristics largely vary from industry to industry?

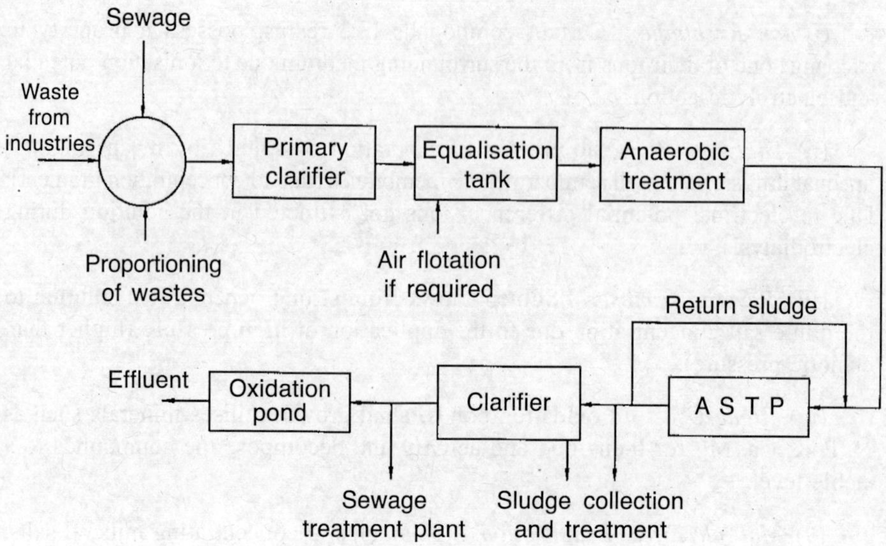

Fig. 5.8 CETP flow diagram.

2. Is it possible to design a treatment plant with such heterogeneous data from different industries?
3. Is the available data reliable? Does this data correlate well with the designed data of CETP effluent?
4. Does biological activity flourish in such a hazardous chemical effluent atmosphere?
5. What is the efficiency expected from a CETP?
6. Whether the effluent from CETP satisfies the norms for ultimate disposal conditions?
7. What is the legal responsibility between the industry and CETP towards the goal of safe disposal of wastes?
8. Who is to monitor the pollution characteristics at each and every stage starting from collection to disposal?
9. Whether relevant data on CETP functioning is subjected to any scientific evaluation?

SUMMARY

- Industrial wastes are very toxic in nature and are not likely to be amenable for simple biological treatment.
- Unless these wastes are properly treated before disposal, either into a public sewer or into large bodies of water, pollutants cause severe environmental degradation.

- Reduction of strength and volume, neutralisation and equalisation are some essential steps before treating industrial wastes.
- Process operations, pollutant characteristics and waste treatment procedures for various industries are briefly outlined in this chapter.

REVIEW QUESTIONS

1. List out the harmful effects due to disposal of industrial wastes without adequate treatment.
2. Argue the merits and demerits for combining industrial and domestic wastes before treatment and disposal.
3. Name the pollutants from fertiliser and tanning factories.
4. Outline the processes available for treating liquid wastes from a pharmaceutical unit.
5. Discuss the utility of aerobic and anaerobic lagoons in the treatment of industrial wastes.
6. Explain the paper manufacturing process, locating the sources of generation of wastes and indicate the lines of treatment.
7. Write short notes on
   (a) characteristics of textile wastes
   (b) removal of chromium
   (c) pretreatment of industrial wastes

OBJECTIVE TYPE QUESTIONS

A. ***Give the most appropriate answer for each question from the four choices provided in each of them.***

1. When Industrial wastes are discharged into inland surface waters, tolerance limit for BOD is prescribed as
   (a) 50 mg/L (b) 40 mg/L
   (c) 30 mg/L (d) 20 mg/L
2. Which industrial waste is likely to have a high BOD?
   (a) Pharmaceuticals (b) Tanneries
   (c) Distilleries (d) Textiles
3. When industrial wastes are discharged into rivers, oxygen depletion takes place due to
   (a) Suspended solids (b) Inorganic solids
   (c) Organic solids (d) Algae
4. In which industry, anaerobic treatment is preferred?
   (a) Dairy (b) Fertiliser
   (c) Pharmaceutical (d) Paper

B. *Fill in the blanks with suitable words:*

1. Partially treated industrial wastes are directly discharged into rivers when _______.
2. Biodegradability of wastes is indicated by _______.
3. The process of mixing acid wastes with lime slurries is known as _______.
4. Activated sludge treatment plant is not recommended for distillery wastes because _______.
5. Dairy wastes may be used for irrigation purposes after _______.

C. *Match the following:*

	Characteristic	Industry	
1.	Chromium	Textiles	( )
2.	Lignin	Tanning	( )
3.	Highly alkaline	Paper	( )
4.	High dissolved solids	Fertilisers	( )
5.	Ammonia	Pharmaceuticals	( )

CHAPTER 6

Air Pollution

6.1 POLLUTANTS: SOURCES AND EFFECTS

6.1.1 Introduction

Clean dry air contains 78.09% nitrogen and 20.94% oxygen by volume. Other gases like carbon dioxide, argon, neon, helium, methane, etc. are present in negligible quantities.

Atmosphere is the column of air that surrounds the earth. The region of atmosphere extending from the surface of the earth upto an altitude of approximately 16 km is known as troposphere. The layer extending from an altitude of 16 km to 49 km is known as stratosphere.

Atmosphere is a dynamic system, which carries a range of chemical substances released from various natural and man made sources. These chemicals react within the atmosphere and finally reach a sink such as the ocean or a receptor such as man. Burning of fuel in vehicles and various process reactions in industries release harmful and toxic chemicals, smoke, particulate matter and gases into the atmosphere. The quantity and composition of impurities depend upon the fuel, raw materials, process reactions, plant capacity and control devices used. Sometimes, when atmospheric conditions are not favourable for wind movement in horizontal or vertical direction, undesirable particulate matter and gaseous impurities accumulate and concentrate in the ambient atmosphere.

Air pollution is caused due to the presence of one or more contaminants in the air in sufficient concentration, composition and characteristics (for such duration) so as to cause injury to life (human, animal and plant), damage to materials and property or to interfere with normal activities, comfort and heritage.

General impurities are organics (methane, benzene, formaldehyde, chlorinated hydrocarbons), inorganics (NO_x, SO_x, CO_x, H_2S, HF, NH_3, Cl_2) and particulate matter (dust, smoke, fumes, droplets).

6.1.2 Sources

Industries, power plants, foundry units, etc. are stationary sources, normally the emissions will not travel beyond a particular area nearby. Vehicular transport systems on road, rail or airways are mobile sources acting mostly along a specific route.

Primary pollutants are SO_2, H_2S, NO_x, CO, HF, NH_3, etc. that enter directly into the atmosphere from the place of generation. Secondary pollutants are O_3, peroxy-acetyl nitrate (PAN), aldehydes and oxidants that are formed in the atmosphere due to the chemical interaction of primary pollutants, especially HC and NO_x with other substances utilising solar energy. Some emissions are produced due to natural events like storms or earthquakes. The processes of industrialisation and urbanisation release high levels of contaminants into the atmosphere, in selected zones.

Major primary pollutants are listed below (Figure 6.1).

Carbon monoxide	—	45%
Oxides of sulphur	—	13%
Particulate matter	—	15%
Oxides of nitrogen	—	10%
Hydrocarbons	—	12%
Others: H_2S, HF, HCl, NH_3, Pb, Hg, Si, etc.	—	5%

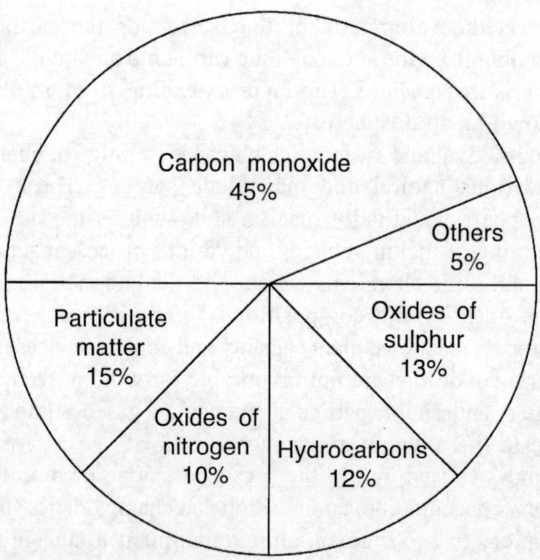

Fig. 6.1 Primary pollutants in the atmosphere.

Natural sources

Volcanic eruptions, forest fires, storms, soil erosion, electrical discharges, oxidation, condensation, polymerisation, photochemical and other reactions in the atmosphere, naturally contribute some contaminants into the ambient air. Oxidation of methane from decaying vegetation results in the formation of CO. Use of fertilisers and insecticides in harvesting and agricultural systems may leave residual chemicals like arsenic, lead, fluorides, etc. in the top soil.

Salt sprays from ocean, air borne particles from soil vegetation, microorganism, spores and pollen grains from agricultural plants and weeds, smoke and CO from fuel burning, CO_2 from respiration and fuel burning, H_2S and CH_4 from anaerobic decomposition of organic wastes, radioactive minerals from the earth's crust, hydrocarbons from trees, are some examples of natural sources of pollution.

Combustion

(a) Combustion of fuel (coal or oil) in automobiles, ships, rail engines and air crafts produces carbon monoxide, smoke, particulate matter, HC, Pb, NO_x, etc. Impurities in the raw material, process of combustion, efficiency of burning, air fuel ratio and ignition temperature decide on the quality, quantity and nature of the pollutants produced. Transportation is a major source of pollution and carbon monoxide is the main pollutant released into the atmosphere, (together with HC and NO_x).

(b) Fuel combustion is also essential in various domestic and industrial units. Fuels containing sulphur as an impurity, release SO_2 on combustion. Oxides of some metals may also be present in small quantities in air due to combustion process. These operations result in smoke, fly ash, particulates, gases, like SO_2, NO_x, CO, etc.

(c) Solid wastes are burnt as a way of disposal.

Process industries

Different air pollutants are released based upon the manufacturing process and the raw materials used. They are detailed in the following table.

Industry	*Impurities*
Cement	Process dust, cement, lime, oxides of Al, K, Si, Ca, Na
Fertiliser	Dust particles of coal, sulphur, gypsum, NO_x, HF, NH_3, F, CO_2, SO_2
Paper	H_2S, mercaptons, SO_2, chemical dusts, Cl_2, methyl sulphide
Thermal power plants	Fly ash, NO_x, SO_2, HC, CO
Iron and steel	Iron oxide, dust, CO_x, cyanides, H_2S
Metallurgical	Fumes, dusts, smoke, oxides
Heavy chemical industry	Acid fumes
Sulphuric acid	SO_2
Nitric acid	NO_2
Petrochemicals and refineries	H_2S, HC, Particulates, CO, HCHO, NH_3, NO_2, catalyst dusts cyanides, etc.

(*Contd.*)

Industry	*Impurities*
Coal tar	Polynuclear hydrocarbons, aerosols of tar
Leather and tanneries	Mercaptons, sulphides
Paints and pigments	Nitrobenzene, aniline, thinners & solvents
Lead casting and melting	Tin, lead fumes, and oxides
Sugar	H_3PO_4, SO_2

Some pollutants and toxic substances are emitted by different industries or chemical processes. These are shown in the table given below.

Impurity	*Process/Industry*
As	Mining, pesticides, chemical wastes
Mn	Mines
Zn	Metal plating
Pb	Storage batteries, combustion of gasoline, paints, smoking
Cd	Metal plating, incineration of plastics, mining
Vd	Petroleum products
Cr	Stainless steel and alloy industries, metal plating, cooling towers
Cobalt	Alloys and tools manufacturing, Hg, Cu, steel, Zn, glass factories
F	Ceramic, glass manufacturing
SO_2	Smelter ore process, sulphuric acid plant
Ni	Combustion of coal or diesel, alloys manufacturing, tobacco smoking
Hg	Combustion of fossil fuels, ore mining, pesticides, vinyl chloride plant, paints, pharmaceuticals, metal smelting
Be	Ceramic, coal, nuclear power, space industry
Asbestos	Fibres from manufacturing units
CH_4	Bio decomposition of matter
HC	From automobiles, air crafts, petroleum refining, power plants

Other sources of pollution

Collision, abrasion and grinding operations release dust and particulate matter. Earth moving, construction work and a number of other engineering activities also cause air pollution.

6.1.3 Characteristics of Air Pollutants

The nature and mechanism of chemical reactions involved with major pollutants are discussed in the following sections.

Particulates

These are minute and discrete particles, capable of settling in gaseous medium. Size, size distribution, mass concentration, chemical composition and features, describe the source of origin, the level of toxicity, the transport behaviour and suggest the mechanism of removal. Dust (1 to 200 μm) is due to mechanical process of disintegration due to grinding, crushing and spraying operations. Dusts may be inert or reactive. They may contain toxic chemicals like arsenic, lead, nickel. Particles of > 1 μm size, settle due to their size and weight. 1 to 10 μm size particles come from soil, dust, soot, emissions from cement and industrial units. Large size particles are from soil erosion, construction and other operations. Small particles (0.01 to 1 μm) are from combustion. Various chemical and metallurgical processes release fumes, usually generated by condensation from gaseous state. All air borne colloidal (solid or liquid) suspensions of less than 1 μm size are called aerosols. They behave aerodynamically as gases and remain in suspension for a long time. Mist (< 10 μm) is dispersion of liquid particles in low concentrations. A complex mixture of gas and fine particles is known as smog. Fog is the mist formation due to visible aerosols in liquid dispersion phase. Visibility in the atmosphere is reduced due to fog.

Different particles behave in different ways. Their particle movement, rate of cooling or evaporation, scattering of light, settling rate, interaction and resistance, adhesion, sorption, migration, diffusion and other properties mostly depend upon the particle size and nature.

Particulates may also be classified according to the following characteristics.

(i) Inorganic dusts (iron oxides, lead halides, oxidation products).

(ii) Droplets (H_2SO_4 and $(NH_4)_2\ SO_4$ mists),

(iii) Fly ash (coal combustion product containing SiO_2, Al_2O_3, Fe_2O_3).

(iv) Organic particulate matter (benzene, polycyclic aromatics),

(v) Smoke (0.03 μm to 1 μm) containing mainly carbon due to incomplete combustion.

Sulphur dioxide (SO_2)

Sulphur is an impurity in coal. Metals are extracted from sulphide ores. Hence SO_2 is mainly released from fuel combustion and metallurgical processing units, as

$$S + O_2 \rightarrow SO_2$$

$$SO_2 + 1/2\ O_2 \rightarrow SO_3$$

$$SO_3 + H_2O \rightarrow H_2SO_4$$

$$FeS_2 + O_2 \rightarrow Fe_2O_3 \text{ and } Fe_3O_4 + SO_2$$

Sulphur dioxide is a product of oxidation in the atmosphere and in the presence of moisture content, forms sulphuric acid droplets. SO_2 is a pungent gas.

Carbon monoxide (CO)

Fuel combustion and oxidation of methane release CO into atmosphere. Soil, vegetation, sea water and microorganism remove CO from the atmosphere.

Public transportation emits CO and CO_2. Various industrial operations like blast furnaces, petroleum refineries, coal mining, etc. also produce CO.

$$2C + O_2 \rightarrow 2CO$$

$$CO_2 \rightarrow CO + O$$

$$CO_2 + C \rightarrow 2CO$$

CO formation depends upon factors like air-fuel ratio, ignition temperature.

Oxides of Nitrogen (NO_x)

NO and NO_2 are produced due to human activities, use of nitric acid in chemical reactions, furnaces and large power plants.

NO is released from fuel combustion and is further oxidised to NO_2. In the presence of sun light many photo-chemical reactions take place forming nitrates and nitric acid. Hydrocarbons interact in the process. Secondary pollutants are more toxic in nature.

$$N_2 + O_2 \Leftrightarrow 2NO$$

$$NO + 1/2O_2 \Leftrightarrow NO_2$$

NO_2 emissions can be reduced by reducing peak combustion temperature (and residence time) and by reducing excess air availability for combustion.

Hydrocarbons

Trees emit large quantities of hydrocarbons. Methane is released from anaerobic reactions and is having very high residence time in the atmosphere. Human activities (including combustion of petroleum products, distribution of petroleum products, etc.) release hydrocarbons in the urban atmosphere. Non-methane group are oxidised photochemically in the presence of sunlight and oxides of nitrogen resulting in O_3, PAN, etc.

Other chemicals

H_2S (a foul smelling gas) comes from anaerobic (biological) degradation of waste matter; kraft pulp industry and petroleum refineries. HF results from aluminium industries and phosphatic fertiliser plants. HCl and Cl_2 are pungent and acidic in nature.

6.1.4 Episodes

Air pollution hazards have been reported from Meuse Valley, Belgium (1930), Los Angeles, California (1944), Donora, USA (1948), Thames Valley (1962), New York (1966) and Tokyo (1970) due to adverse atmospheric conditions. Bhopal, India (1984), Chernobyl (1986), Mexico (1952) and Mississippi (1969) episodes are due to accidental release of pollutants. All these incidents brought suffering and death to the local population on a large scale.

The hazards due to air pollution assume a serious proportion when

- A number of factories are located at one place
- Industries are located in a valley allowing less dispersion and dilution
- Pollutants are continuously released into the atmosphere
- Pollutants accumulate at the ambient level
- Fog (moisture content) is present in the atmosphere
- Wind movement and dispersion of pollutants are minimum
- Temperature inversion conditions occur in the atmosphere

In addition, hazardous secondary pollutants are also created in the atmosphere due to photochemical reactions.

It was found that SO_2 and particulate impurities were present in higher concentrations at ground level during Belgium, Los Angeles, Donora and London incidents, which resulted in unexpected sickness and death of many people in few days. Release of H_2S caused severe damages in Mexico and Mississippi. Pesticide factory at Bhopal, accidentally released toxic methyl isocyanide, killing and injuring lakhs of people and causing severe property damages. In Chernobyl, there was a leakage of radioactive material into the atmosphere that caused severe damages to life and property. Even the famous Taj Mahal (India) is being affected by the air pollutants being released continuously from the nearby oil refineries, foundries and other industries. These revelations are sufficient to prove the gravity of the situation when air pollutants accumulate at the ground level for a long duration under worst atmospheric conditions.

6.1.5 Harmful Effects

The highly toxic and poisonous emissions have extremely harmful effects on human health and the environment which may be divided into the following categories. Air pollution may cause

- Material damage including soiling of surfaces
- Vegetation damage
- Reduction in visibility and darkening of the sky
- Damage to human health and domestic animals
- Interference with services and production
- Annoyance and psychological disturbance

The damage depends upon the nature of the pollutant, concentration of the pollutant, duration of exposure and individual susceptibility. Atmospheric conditions like temperature, humidity also are important.

Air pollution causes damage mainly of the respiratory tract. Some chemicals also enter into the nervous system while some damage kidneys, heart and liver. Irritation of the skin, eyes, nose, throat, cough, nausea, suffocation, breathlessness, allergy are some common symptoms. Bronchitis, Asthma, dermatitis may occur depending upon the chemicals present in the air.

While exposure to higher concentrations of hazardous gases/emissions for a short duration causes acute conditions (as learnt from the disasters that occurred in the past), continuous exposure to lower concentrations also results in chronic diseases (as observed in the case of workers in a factory). Aerosols also affect weather and damage material and human health due to the presence of toxic chemicals.

Human beings. Carbon monoxide has high affinity towards haemoglobin and reduces oxygen carrying capacity of blood, leading to damage of central nervous system. Asphyxiation is caused by CO_2 while SO_2 causes irritation of the respiratory tract and cough, and NO_2 (more toxic than NO) causes inflammation of lungs. Particulate matter accumulates in the lungs of those working in coal mines, asbestos units, etc. causing irreparable damage. Pollen, spores, animal fur and some fine powders are known as aero allergens. Beryllium, nickel, benzopyrene are carcinogenic. Cadmium (Cd) causes cardiovascular diseases, and damage to kidney and liver, and Mn acts on nervous system. Lead (Pb) is neurotoxic and mercury (Hg) causes damage to brain and kidney. Fluoride (F) compounds act on bones while arsenic (As) is a poison. Some of the hydrocarbons (alkenes, ethylene, etc.) are highly reactive and form toxic secondary products. Cr, Vd, F, As, Pb and other toxic chemicals are also injurious to the livestock and animals.

Vegetation. Dusts and aerosols are deposited on leaves, affecting the rate of photosynthesis and evapotranspiration. This results into premature dropping of leaves (abscission), reduction of chlorophyll (chlorosis) and leaf tissue damage (necrosis), which ultimately lead to the plant decay. Fluorine and O_3 affect cell metabolism, and SO_2 acts on chlorophyll, cells and tissues. All these hamper the process of photosynthesis and growth in plants.

Plants are more sensitive when compared to animals and human beings. Arsenic and fluorides on leaves are poisonous to animals consuming the plants. Peroxy-acetyl nitrate (PAN) and O_3 affects photosynthesis and oxidants inhibit the enzyme activity. NH_3 and NO_3 are also injurious to plants.

Materials and property. Air pollution has been a significant source of economic loss in urban areas due to damage to materials in five ways: (i) corrosion, (ii) direct chemical attack, (iii) indirect chemical attack, (iv) abrasion and (v) deposition.

Dust, soot and fumes react with painted surfaces, fabrics, buildings, sculptures and metals. Even in small concentrations, SO_2 leads to corrosion of metals (steel and aluminium), and acts on paints, textiles, paper, leather goods, leading to discolouration and disintegration of the material. Sulphuric acid mist reacts with carbonate building materials like marble and limestone. NO_x acts on fibres, textiles and dyes. Ozone acts on rubber and fabrics. Oxidation results in fading of colour and material damage. Acid gases react with ceramic products.

Other environmental factors such as moisture, air-salt content, temperature, sunlight, etc. also play a crucial role in reducing the strength, quality, utility and value of the materials and property.

Visibility. Poor visibility is an indicator of air pollution. Visibility depends on the transmission of light through the atmosphere. Pollutants scatter and absorb light. Visibility gets reduced as fog formation traps acid aerosols and particulate matter in the air. Reduction depends upon the size, concentration and physical characteristics of particulate pollutants present in the atmosphere. Under humid conditions, some particles pick up moisture and increase in size and affect the visibility.

6.1.6 Atmospheric Reactions

Acid rain. In the presence of moisture, SO_2 forms sulphuric acid aerosol droplets in the atmosphere. Similarly, NO_x forms nitric acid. These two acid suspensions join water sources during the rainy season bringing the pH of soil and water to less than 5.0. Acid rain causes serious ecological and irreversible changes in the environmental conditions, since it affects soil and vegetation and reduces fish and aquatic life. Acids also react with building materials.

Inversion. Normally atmospheric temperature is higher at a lower elevation, i.e. nearer to the ground. Although air rises and moves up vertically when it gets warm, but sometimes a warm dense layer acts as a lid in the atmosphere and prevents the upward movement of air, (including its contaminants) from ground level. The condition of cool and dense air nearer to the ground and warm air at higher elevation is known as inversion condition. This results in the concentration and accumulation of harmful chemicals at the ground level. In a calm and quiescent wind condition, (and if fog forms) there will be little mixing and less dispersal of pollutants. In that case, life and materials in that particular locality suffer more due to the accumulation of various chemicals.

Green house effect. Nowadays more CO_2 is released due to high rate of fuel combustion. Because of the fast depletion of forests, CO_2 level in the atmosphere is rapidly increasing. Since water vapour and CO_2 trap a large fraction of earth's emitted radiation near the ground level, the earth gets heated up slowly. Atmospheric temperature is also altered due to the presence of dust and particulate matter. Green house effect (Fig. 6.2a) means heat is trapped near ground level increasing global temperatures by few degrees. This may result in melting of ice, increase in sea level and finally leading to submergence of coastal areas. Higher temperatures also affect seasons and crops. Reducing the CO_2 emissions in the atmosphere and developing programmes like afforestation to utilise CO_2 are a few remedial measures that can protect the environment.

Ozone layer. Ozone in the ambient atmosphere is responsible for a number of secondary chemical reactions, the end products being more harmful to human beings, plants, animal life and materials. Trace quantities of ozone in stratosphere, protects life on earth from the deadly ultraviolet radiation. Industries release CFC, chlorinated and brominated chemical compounds into the atmosphere and destroy the protective ozone layer as chlorine reacts with ozone (Fig. 6.2b). Ozone depletion means more exposure of solar radiation causing damage to human ecosystems. Food production is reduced. Human beings are exposed to skin cancer, blindness and more serious hazards.

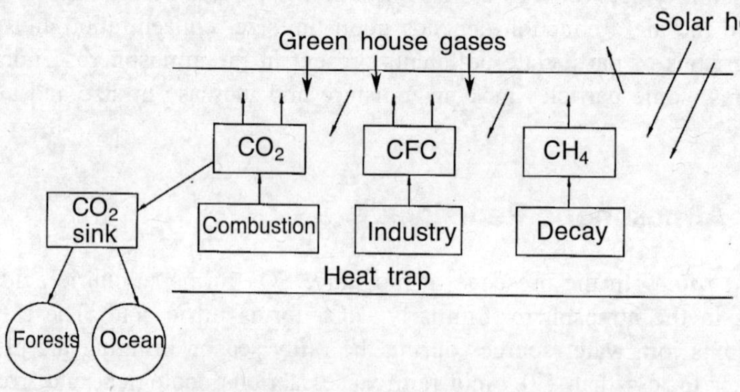

Fig. 6.2(a) Green house effect.

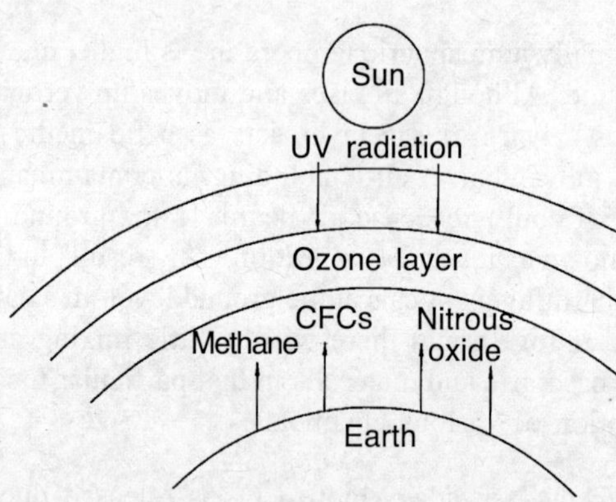

Fig. 6.2(b) Ozone layer.

Photochemical smog. Pollutants such as HC and NO_x in the presence of sunlight and ozone (or active 'O') forms different compounds such as formaldehyde, acrolein, acrosol and peroxy-acetyl nitrate (PAN). Formation of photochemical smog reduces visibility, causes irritation, cracking of rubber and fading of dyes and damages plants and vegetation. Chemical reactions involved in the formation of smog and PAN formation are presented briefly.

(i) Formation of sulphuric acid aerosols:

$$SO_2 + O_2 \rightarrow SO_4$$

$$SO_4 \rightarrow SO_3 + O$$

$$SO_3 + H_2O \rightarrow H_2SO_4$$

(ii) Formation of nitric acid:

$$NO_2 \rightarrow NO + O$$

$$O + O_2 \rightarrow O_3$$

$$NO + O_3 \rightarrow NO_2 + O_2$$

$$O_3 + NO_2 \rightarrow NO_3 + O_2$$

$$NO + NO_2 \xrightarrow{\text{in the presence of } H_2O} HNO_2$$

$$NO_3 + NO_2 \xrightarrow{\text{in the presence of } H_2O} HNO_3$$

(iii) Formation of peroxy-acetyl nitrate (PAN) in the presence of ozone, hydrocarbons and nitrogen oxides:

$$O_3 \rightarrow O_2 + O$$

$$HC + O \rightarrow HCO \text{ or } RO_2 \text{ alkyl peroxyradicals}$$

$$N_2 + O_2 \rightarrow 2NO$$

$$NO + 1/2O_2 \rightarrow NO_2$$

Conversion of hydroperoxyl radicals to hydroxyl radicals takes place as shown below.

$$HO_2 + NO \rightarrow HO + NO_2$$

Conversion of alkyl-peroxyl radicals to alkyl oxyl radicals takes place in the following manner:

$$RO_2 + NO \rightarrow RO + NO_2$$

$$RO_2 + O_2 + NO_2 \rightarrow PAN$$

Automobile exhaust. Movement of vehicles releases toxic gases like CO_x, NO_x and various hydrocarbons from their exhaust emissions. Smoke and particles of unburnt carbon are also released. Diesel engines release more of smoke and particulate matter whereas gasolene engines release high percentage of hydrocarbons, carbon monoxide and nitrogen oxides. Lead is also emitted by the exhaust of the petrol engines.

A number of factors, viz. air-fuel ratio, combustion mechanism, ignition temperature, exhaust system, spark timing, engine condition, fuel adulteration, and vehicle maintenance, play important roles in producing various pollutants from automobile exhaust. Figure 6.3 shows the effect of air-fuel ratio on the production of various emissions. Depending upon traffic conditions, CO level on the road may be around 40 ppm. Dust is also released depending upon the road and traffic conditions. These pollutants cause damage to life and property as discussed earlier.

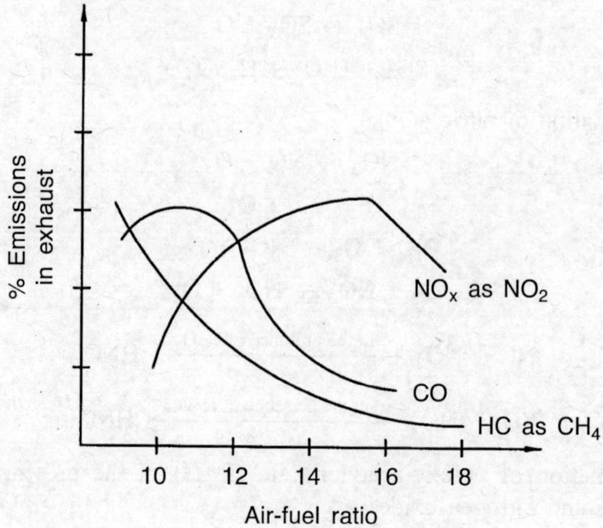

Fig. 6.3 Automobile emissions.

6.1.7 Air Quality Permissible Limits

	Pollutant	*Permissible level* (μg/m^3)
(i)	Suspended particulate matter	75 (annual geometric mean) 260 (24 hours averaging time)
(ii)	Carbon monoxide	10,000 ppm (8 hours) 40,000 ppm (1 hour)
(iii)	Sulphur dioxide	80 (annual average) 365 (24 hours)
(iv)	Nitrogen dioxide	100 (annual mean)
(v)	Ozone	235 (1 hour)
(vi)	Lead	1.5 (90 days average)
(vii)	Photo chemical oxidants	160 (1 hour)
(viii)	H_2S	42 (1 hour)
(ix)	Vinyl chloride	26 (24 hour)

Notes. (a) All these pollutants are measured at ambient level. However, care should be taken first to control emission levels at source.

(b) Peak levels and average levels are both important.

(c) Permissible values for an industrial area are different from that of the residential locality.

6.2 DISPERSION OF POLLUTANTS

6.2.1 Atmospheric Factors

Air quality depends upon topographical and meteorological conditions. Weather of a particular place, atmospheric temperature, pressure and humidity, rainfall, local winds, etc. constitute the local climate. It is influenced by latitude and altitude of the place, its proximity to sea, presence of mountains, land and water surfaces, forests, surface air movements, earth's position with respect to the sun, and other parameters. Human activities in a locality depend upon industries, urban development, deforestation, use of surface and ground waters, use of heat and energy sources and transport systems, all of which also influence the local environmental conditions. Whether pollutants are deposited or carried away, (settle or accumulate) and the mechanism of dispersal (transport, react or deposit) mainly depend on mixing conditions, slow or strong winds, coastal conditions, valleys, inversion and location of industries with reference to wind-flow direction. All these atmospheric factors affect the pollution level at a particular place.

6.2.2 Air Movement

Earth's rotation and revolution around the sun cause unequal distribution of atmospheric temperature and pressure due to non-uniform heating of places on the earth. Local factors (geographical, meteorological, topographical) including the climate, decide on the horizontal movement of air, while temperature, density and atmospheric stability control the air movement in the vertical direction. Hot air rises up at the equator and cool air from poles moves towards the equator at a lower level. West to east rotation of the earth deflects the winds right or left in northern or southern hemisphere.

Warm air expands and moves to adjacent regions of lower temperature. This results in the reduction of weight (density) of warm air, and increase of density (weight) of surrounding cooler air. Then the cooler (heavier) air mass descends down and gets compressed, whereas the warmer air moves upwards. When the atmospheric temperature is high, or during day time, air movement is rapid. Wind speed, turbulence and unstable conditions occur near to the surface of the earth. Stable atmosphere is not desirable as the dispersal of pollutants in the ambient atmosphere is affected.

Earth's surface terrain roughness affects wind velocity upto 1 km above ground level.

The following changes are generally observed in respect of air flow and air temperature due to variations in parameters such as altitude, height, temperature and pressure of the land area.

1. Atmospheric turbulence is due to eddies and short circuiting in the environment as shown in Fig. 6.4 from obstructions.
2. At ground level, air moves over buildings, trees, rocks, etc. which create

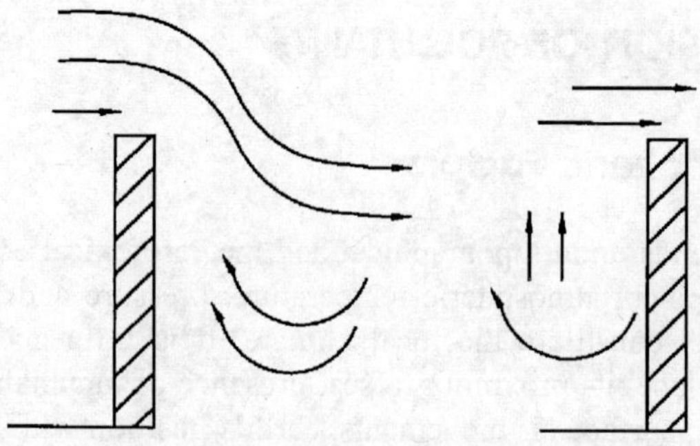

Fig. 6.4 Air movement (Obstructions).

friction and turbulence. Urban road surfaces, open areas, also create radiation heat at ground level.

3. Air near the mountain slope warms faster during day time than at the same altitude but away from the mountain. This causes a local pressure gradient towards the mountain and air is forced to flow up the slope as shown in Fig. 6.5. The reverse happens during the night time.

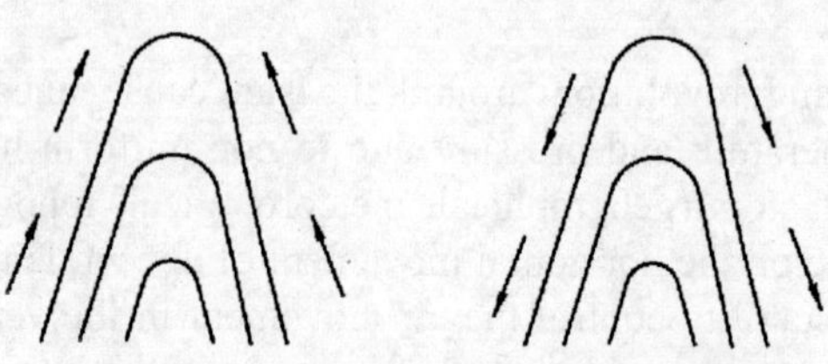

Fig. 6.5 Air movement (Mountain-valley).

4. Plumes discharging pollutants in a valley may confine them in the valley only due to restricted horizontal movement.
5. Sea breeze occurs in the day time and land breeze blows at night in coastal places. Land is heated up rapidly during day time. As a result, upper layer of air moves from land to sea and low level air moves from sea to land. Land air cools more rapidly during night whereas air temperature changes rather slowly over water bodies. This causes land and sea breezes as shown in Fig. 6.6. Pollutants get accumulated on land in coastal areas during night times. Cool and heavy air over a warm area creates unstable conditions.

All these factors decide on the movement of wind at a given place at a given time. It is possible to locate an industry away from residential zones if the wind direction and velocity of that place are known. Wind rose diagram (Fig. 6.7) illustrates the air movement at a place. Dispersion pattern of pollutants are known from this diagram. Eight directions are marked on the diagram to represent wind

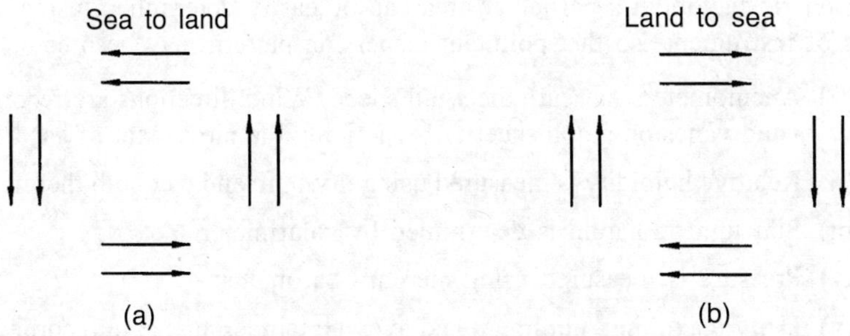

Fig. 6.6 Air movement (Land-sea).

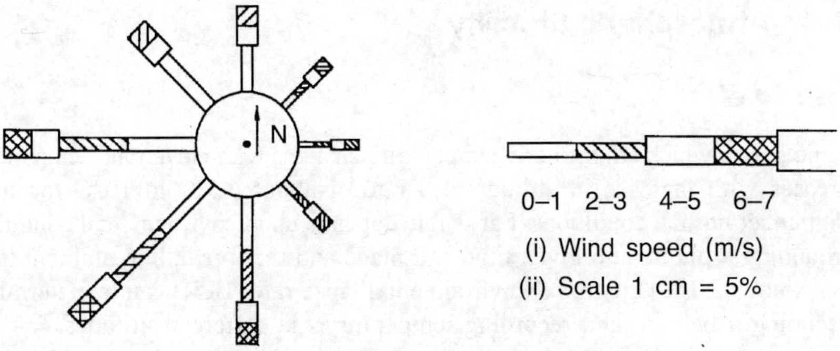

Fig. 6.7 Wind rose diagram.

movement in that direction. Wind speed is represented on the line to some scale (in that direction) and also the frequency of occurrence is marked on the wind rose.

6.2.3 Heat Transfer

Earth absorbs solar energy and reflects the same back into space. A number of factors affect this phenomena. Sun's rays are absorbed by water, CO_2, O_3, dust and other particulate matter. Heat transfer by direct physical contact of air (earth) is called conduction. It depends on the distance from the surface of earth. Rising of warm air and sinking of cold air is a process of convection. Land, water, and air do not absorb or transmit heat at the same rate. This unequal heating and cooling causes variation of air temperatures. Also, temperature decreases by about 1°C for every 100 m increase in the altitude (adiabatic lapse rate). Negative lapse rate causes temperature inversion.

Humidity

Water vapour present in the air affects the rate of cooling, absorbs radiation, stores latent energy, changes air temperature and air density. Particulate or gaseous pollutants may be retained due to the humid conditions in the atmosphere. All the atmospheric factors at a particular area can be easily determined, using various types of instruments so that pollutant dispersion pattern may be predicted.

(i) Anemometers measure the wind speed. Wind directions are recorded by wind vanes on graph sheet or by noticing the movement of gas balloons

(ii) Relative humidity is measured using dry bulb and wet bulb thermometers

(iii) Sun light radiation is determined by solarimeter

(iv) Pressure is measured using mercury barometer

(v) Daily maximum, minimum and average temperatures are recorded using different types of thermometers.

6.2.4 Atmospheric Stability

Lapse rate

It is the rate at which atmospheric temperature changes with elevation. Temperature decreases with increase in altitude at a rate of about 6.5°C to 10°C for 1 km height under normal conditions. Lapse rate depends upon environmental conditions surrounding a plume at a given time and place and that prevailing ambient lapse rate is known ELR (Fig. 6.8). Environmental lapse rate (ELR) can be determined by sending a balloon and recording temperatures at different altitudes

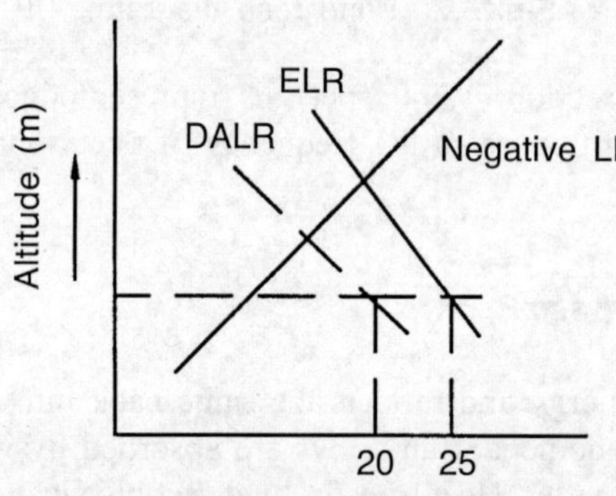

Fig. 6.8 Lapse rate.

Dry adiabatic lapse rate (DALR) is measured under dry or normal conditions, whereas wet adiabatic lapse rate (WALR) is taken in the presence of moisture (condensation of water vapour releasing latent heat). It has been observed that WALR (6°C/km) is more than DALR (9.8°C /km).

When temperature increases with altitude, it is known as negative lapse rate. Altitude also affects atmospheric pressure, density and other gas properties. If the temperature of a certain mass of air is greater than that of the ambient air, the parcel of the warm air rises and expands until its own density is equal to that of the surrounding atmosphere. Then it becomes lighter and cooler. As per the laws of thermodynamics, temperature decreases as the internal energy decreases. Change in internal energy, specific heat, work done on the surroundings, result due to variation of air temperature with altitude. *Adiabatic lapse rate* (ALR) can also be theoretically evaluated at different altitudes. Vertical movement of air depends upon the lapse rate under different environmental conditions.

Stability

If a rising parcel of air remains warmer than the surrounding air, or if a descending parcel remains cooler, then the displaced air quantity continually moves away from its original position, carrying the pollutants further away, and favouring dilution and dispersion. This is known as unstable condition.

If a rising parcel of air arrives at an altitude in a cool or denser state than the surrounding air, then it results in air moving downwards towards the earth, i.e. away from the direction of the desired displacement. This is known as a stable condition. That means air movement is restricted and pollutants concentrate at a place. If there is no movement upwards or downwards, then it is known as neutral condition.

Stability of atmosphere is its behaviour to prevent the vertical motion of air or suppress causes of turbulence. Air stability is mostly decided by the temperature gradient in the vertical direction (i.e. lapse rate). If the environmental lapse rate (ELR) in the atmosphere is less than the normal adiabatic lapse rate, stability occurs. If the ELR is more than the normal lapse rate, i.e. in super adiabatic conditions (SALR), unstability conditions arise. Stable inversion appears if the temperature increases with higher altitudes. Negative lapse rate or inversion is likely to occur on clearer nights with less wind movement.

A comparison is made between ELR and ALR to understand atmospheric stability (Fig. 6.9) under varying conditions:

(i) Isothermal condition means air temperature is constant throughout a layer of atmosphere (i.e. no change in temperature with altitude). Ambient lapse rate is zero.

(ii) When DALR is equal to ELR under neutrally stable atmosphere, rising parcel of air will not feel any buoyant force. It will have the same pressure, temperature and density as that of its surroundings. So displaced mass of air neither tends to return nor tends to continue its displacement. There is no upward or downward movements.

(iii) When ELR is more than DALR under super adiabatic condition in the atmosphere, rising air (cooling at adiabatic rate) becomes warmer and less dense than the surrounding environment, with more buoyancy, it tends to move upwards. Air from different altitudes mixes thoroughly enhancing favourable dispersion in the atmosphere.

(iv) When ELR is less than DALR under sub-adiabatic conditions rising air parcel becomes cooler and more dense than the surroundings. It tends to fall back to its original position in stable atmosphere. Because of little vertical mixing, pollutants concentrate more rapidly in that environment.

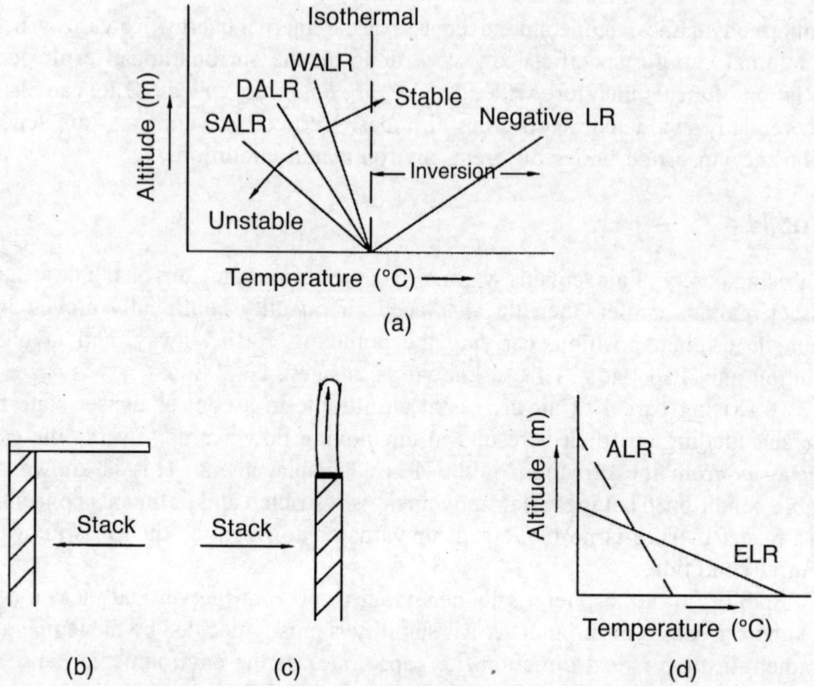

Fig. 6.9 (a) Atmospheric stability, (b) Isothermal condition, (c) Neutrally stable condition, (d) Super adiabatic condition.

(v) When temperature increases with altitude (negative lapse rate), there will be no vertical mixing. It is a highly stable atmosphere under inversion conditions. Warmer air blankets cover colder air pockets. More pollutants are at ground level only.

Horizontal movement of wind is responsible for the reduction of pollutant concentration with respect to time and distance from the source of emission in different directions.

Inversion

This is an unfavourable situation for dispersion of pollutants when temperature increases with altitude (Fig. 6.10). Inversion conditions occur due to various factors, and can be of three different types.

Radiation inversion. After sunset, ground cools quickly by radiational heat transfer. Air in contact with the ground surface loses sensible heat through convection. Near ground level, because of unequal cooling rates of earth's surface as well as the air immediately above, temperature inversion occurs between cool and warm layers of air. This usually happens during nights and breaks with sunrise as sun's heat overturns layers by convective currents. Early morning fog, traps gaseous and particulate pollutants. Valleys and low lying areas are especially affected by these inversions, as dense and cooler air stays below the warm air.

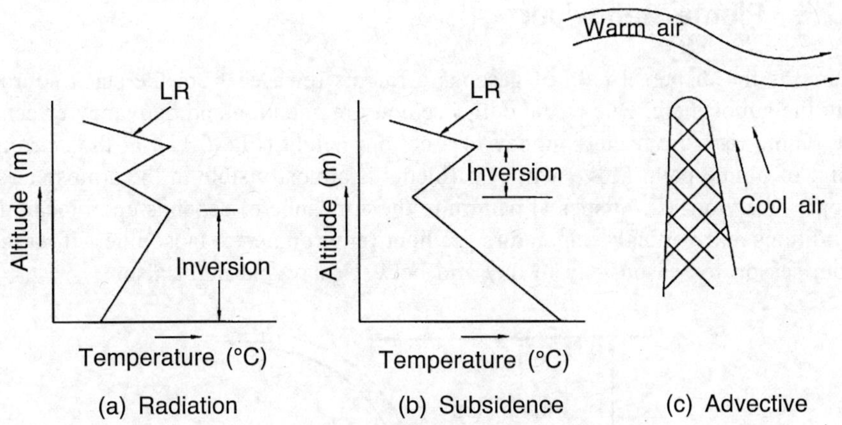

Fig. 6.10 Inversion.

Subsidence inversion. It is caused by air moving down to replace air which has flown out from the high pressure region (anti cyclone). Air is warmed up by compression as it descends in a high pressure system and achieves temperature higher than that of the air below. Blanket of warm air over the cooler air below, prevents the vertical movement of air.

Advective inversion. Hill range forces warm air from land breeze to flow at higher levels and a cool sea breeze to flow at lower levels in the opposite direction. This results in warm air moving over a cool air.

6.2.5 Mixing Depth

Pollutants released from industrial stacks, get mixed up in the air between 100 to 500 m height above ground level, depending upon the meteorological factors prevailing in the locality. This vertical distance where pollutants are diluted and dispersed in the available atmospheric conditions is known as mixing depth. This height depends upon environmental lapse rate at a given place and time as shown in Fig. 6.11. Plume height should not be lower than the mixing depth as it results in ground level concentration of the pollutants.

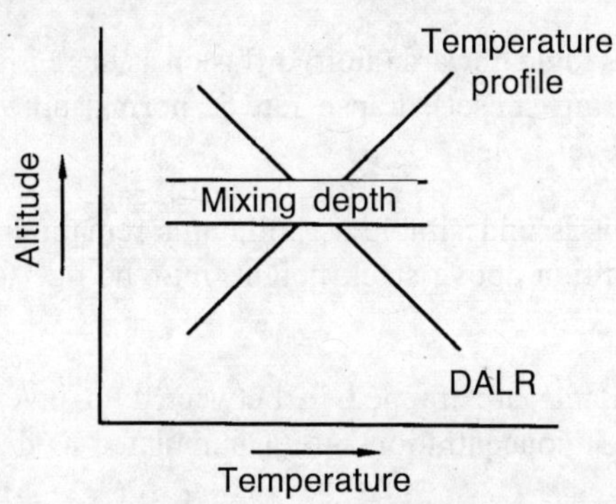

Fig. 6.11 Mixing depth.

6.2.6 Plume Behaviour

Plume is the dispersal path of gaseous effluents released from the stack source into the atmosphere. Due to the outlet velocity momentum and buoyancy effects, the plume rises from the chimney to a certain height (Fig. 6.12), at the starting point of plume path. Movement of effluents is almost visible in the atmosphere. Plume behaviour (i.e. dispersal pattern in the atmosphere) depends upon local air conditions and vertical temperature gradient (environmental lapse rate). It varies from season to season, day to day and between day and night also.

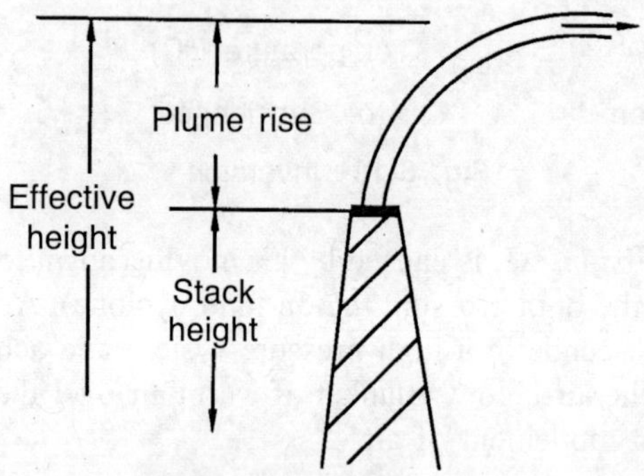

Fig. 6.12 Plume rise.

Seven categories of plume patterns are shown in Fig. 6.13 and their characteristics are described here.

Looping. This is good for the dispersal of pollutants due to the formation of a very unstable atmosphere. It has a wave pattern and provides high degree of mixing at lower levels.

Coning. Pollutants move farther in windy conditions. The lapse rate is weak.

Fanning. Pollutants move slowly in the horizontal direction under isothermal lapse rate, during nights when wind movement is low.

Lofting. It causes rapid upward (aloft) diffusion as there is a radiation inversion below (may form during sunset). Lapse rate is normal above stack height. So pollution at ground level is less.

Fumigation. This is undesirable as pollutants remain at ground level. It is due to inversion condition above stack height, may be due to sea breeze in the forenoon.

Trapping. Pollutants are trapped in between two inversions due to very stable conditions. High concentrations are accumulated as dispersion is limited.

Neutral. The plume rises up into the atmosphere when both ambient and

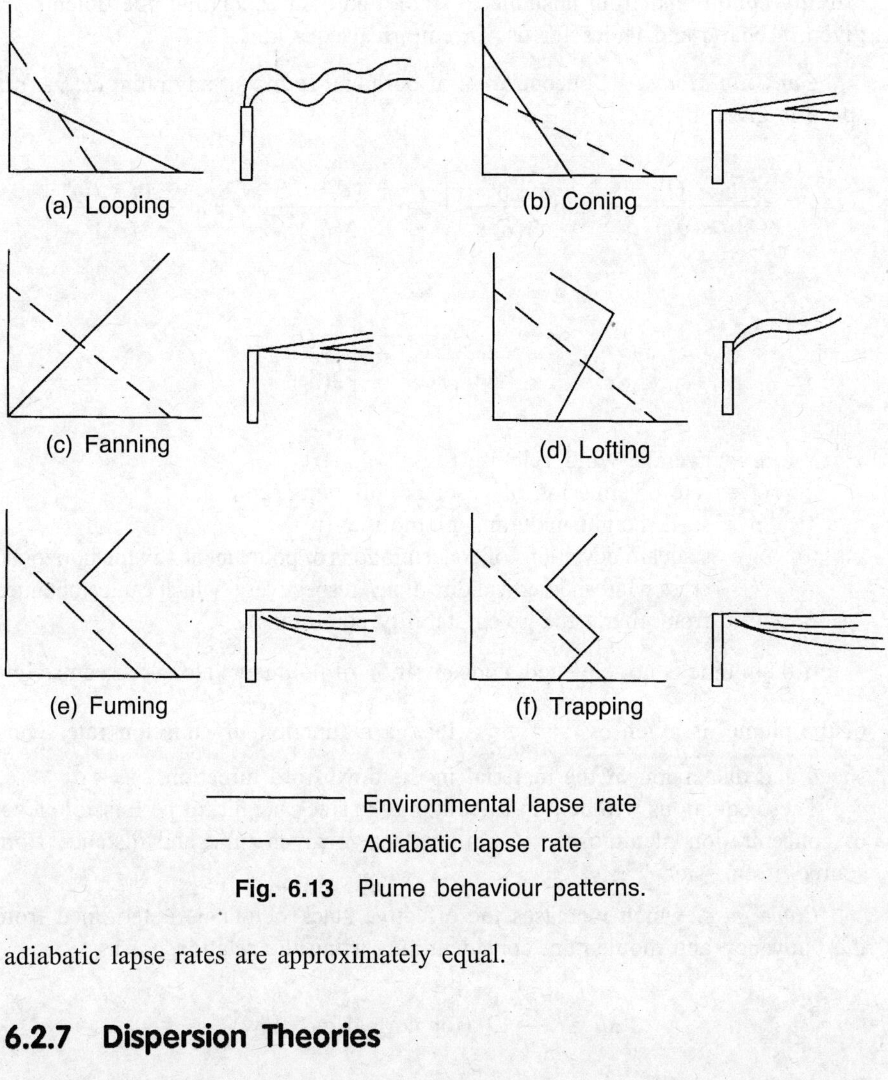

Fig. 6.13 Plume behaviour patterns.

adiabatic lapse rates are approximately equal.

6.2.7 Dispersion Theories

Pollutants disperse both horizontally and vertically away from the stack and get diluted in the atmosphere. Stack height plays an important role. Concentration of pollutants decrease with the lapse of time and increase in distance from the source in the direction of wind movement.

Various sets of empirical equations and theories of diffusion are available to predict the pollution levels with distance and time from the source of emission. Stack height is determined from the wind speed and rate of emissions, so as to keep the pollution within the permissible level at a particular place and time.

Information regarding gas temperature, pressure, concentration and characteristics of pollutant, composition of gaseous mixture, particle size and distribution, exit velocity, wind velocity, direction and turbulence, mean mixing depth, vertical temperature profile of atmosphere, and air stability considerations are utilised in the air diffusion equations to determine the stack height. Atmospheric stability conditions (from unstable to stable) and corresponding coefficients are given in charts and tables for use in empirical equations.

Gaussian Model. Concentration of pollutant (g/m^3) at any point (x, y, z) in space is given by:

$$C_x = \frac{Q}{2\,\pi u(\delta_4)(\delta_2)} \exp \frac{-y^2}{2(\delta_4)^2} \left[\exp \frac{-(z-H)^2}{2(\delta_z)^2} + \exp \frac{-(z+H)^2}{2(\delta_z)^2} \right]$$

or

$$C_{x,0} = \frac{Q}{2\pi u(\delta_y)(\delta_z)} \exp \left[\frac{-H^2}{2(\delta_x)^2} \right]$$

where

u = average wind velocity (m/s)
Q = rate of emission of gaseous pollutants (g/s)
H = stack height including plume rise (m).
δ_4, δ_2 = standard deviations of concentrations of pollutant at x in the horizontal cross wind and vertical directions, respectively (which can be obtained from charts for given stability conditions).

If the plume is not elevated, concentration of pollutants along the centre line of the plume, is given by $\dfrac{Q}{2\,\pi u \delta_y \delta_z}$ that is a function of emission rate, wind speed and dispersion of the material in the downward direction.

These equations will help in calculating the stack height for permissible level of concentration of atmospheric pollutants at a given time and distance from source of emission.

Plume rise, which increases the effective stack height, is determined from the buoyancy and momentum considerations using the relationship as:

$$\Delta h = K \frac{V_s}{u} D \text{ (for normal gas flow)}$$

where

Δh = plume rise (m)
V_s = stack gas velocity (m/s)
u = velocity of wind (m/s)
D = inner diameter of the stack (m)
K = a coefficient based on environmental conditions.

Stack height is given by

(i) $H = 74(Q_p)^{0.27}$ for particulate emissions (Q_p in tonne/h)

(ii) $H = 14(Q_s)^{1/3}$ for SO_2 emissions (Q_s in kg/h)

PROBLEM 1 Determine the stack height required when SO_2 emissions are 100×10^6 g/s and SO_2 concentration of 30 μg/m^3 are permitted at a distance of

500 m from the stack. Assume wind velocity of 5 m/s and stability coefficients as 35 and 19.

Soluton Height of the stack is obtained from using the formula

$$30 = \frac{100 \times 10^6}{2 \times \dfrac{22}{7} \times 5 \times 35 \times 19} \times \exp\left[-\frac{1}{2}(H/19)^2\right]$$

or,

$$H = 45.6 \text{ m}$$

PROBLEM 2 Determine the stack height when the particulate emissions are at the rate of 3×10^{-3} tonne/h

Solution Using the formula, $H = 74(Q_p)^{0.27}$, we have

$$\text{Stack height} = 74(3 \times 10^{-3})^{0.27} = 15.4 \text{ m}$$

(but the minimum stack height should be 30 m)

6.3 REMOVAL OF PARTICULATE MATTER

6.3.1 General

Air pollution may be minimised by:

- selecting proper raw materials
- suitable changes in manufacturing process reactions
- modifying the equipment used
- changing the operational schedules
- recovering or recycling the wastes

Smoke and sulphur dioxide can be reduced by selecting low volatile low sulphur coal. Double catalysis-double absorption process reduces SO_2 emissions in the production of sulphuric acid. Use of proper combustion devices, maintaining proper air-fuel ratio, ignition temperature, stack design and maintenance, reduce the output of pollutants. All these attempts are successful to a reasonable extent. Particles vary in size, shape, distribution and characteristics. The choice of equipment depends upon a number of factors such as

(a) Physical and chemical properties of the particles

(b) Particulate size, concentration and volume of impurities present

(c) Temperature and humidity of the medium

(d) Carrier gas volume and rate of removal

(e) Efficiency required

(f) Economic considerations

Gravity settling chambers, cyclone separators, wet scrubbers, filters and electrostatic precipitators are used for the removal of particulate matter in the industry. Most of them are patent units of manufacturing firms and industries. Various collection devices are briefly described and illustrated to present a basic knowledge of the process and working of each unit.

6.3.2 Settling Chambers

These are mainly used for the removal of larger particles (> 40 μm) or as a preliminary treatment unit before other devices. Air flows at a low velocity (< 100 cm/s), giving sufficient opportunity for settling of particles due to gravity forces only. A particle settles depending upon the horizontal flow velocity, upward forces, resistance due to buoyancy and downward gravitational force on the particle. The more the particle size, the more will be the settling rate.

Settling velocity is given by *Stoke's law* as

$$V = \frac{g d_p^2 (\rho_p - \rho_a)}{18\,\mu}$$

where

g = gravitational force constant (m/s^2)
d_p = diameter of the particle (μm)
ρ_a, ρ_p = densities of air and particles (kg/m^3)
μ = viscosity of air (kg/m · s)

PROBLEM 3 Obtain the settling velocity of particulates, given that diameter of the particle = 1 μm.

$(\rho_p - \rho_a) = 1 \times 10^3$ kg/m^3

Viscosity of air = 2.0×10^{-5} kg/m · s

Solution Settling velocity is obtained by using the formula

$$V = \frac{g d_p^2 (\rho_p - \rho_a)}{18\,\mu} = \frac{9.81 \times (1 \times 10^{-6})^2 \times 1 \times 10^3}{18 \times 2.0 \times 10^{-5}} = 0.28 \text{ m/s}$$

PROBLEM 4 Find the diameter of the particles that will be removed from a settling chamber, given that

V (settling velocity) = 0.33 m/s, $\mu = 2.0 \times 10^{-5}$ kg/m · s,
Sp. gr. of the particle = 2.0, L(Settling chamber length) = 8 m
H(height) = 2 m $\rho_p = 2000$ kg/m^3

Solution Minimum size of particles that are removed, is obtained using the formula

$$d_p = \left(\frac{18\,\mu VH}{gL\rho_p}\right)^{1/2} = \left[\frac{18 \times 2 \times 10^{-5} \times 0.33 \times 2}{9.81 \times 8 \times 2000}\right]^{1/2} = 40\ \mu\text{m}$$

Settling chambers are designed based upon the minimum size of particles to be settled and choosing such a flow rate and surface area loading, to achieve the objectives effectively. Dust removal may be continuous or at different intervals of time. Collection trays improve the efficiency of collection due to the provision

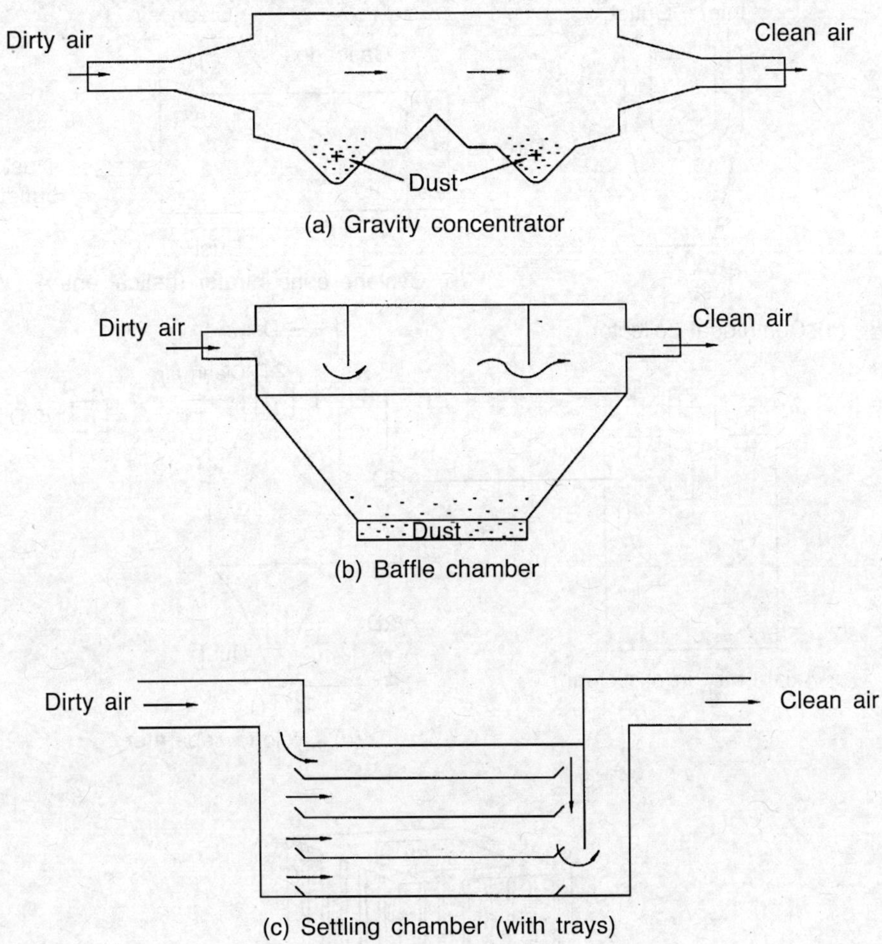

Fig. 6.14 Settling units.

of more surface area for the given flow rate, but dust removal from the unit is difficult. A sudden reduction in velocity helps in a better removal condition in a baffle chamber. The advantages are: low initial and maintenance costs and dry collection of particles. Different types of settling units are shown in Fig. 6.14.

6.3.3 Cyclones

Rotating motion of gas stream generates a centrifugal force and throws the particle outside radially, towards the wall. Gas moves up from inner portion of the cyclone chamber. Particles slide down the cyclone and are collected.

Centrifugal force or the particle separation factor depends upon the inlet velocity (V) and radius (r) of the cyclone as V^2/gr. Efficiency increases when the

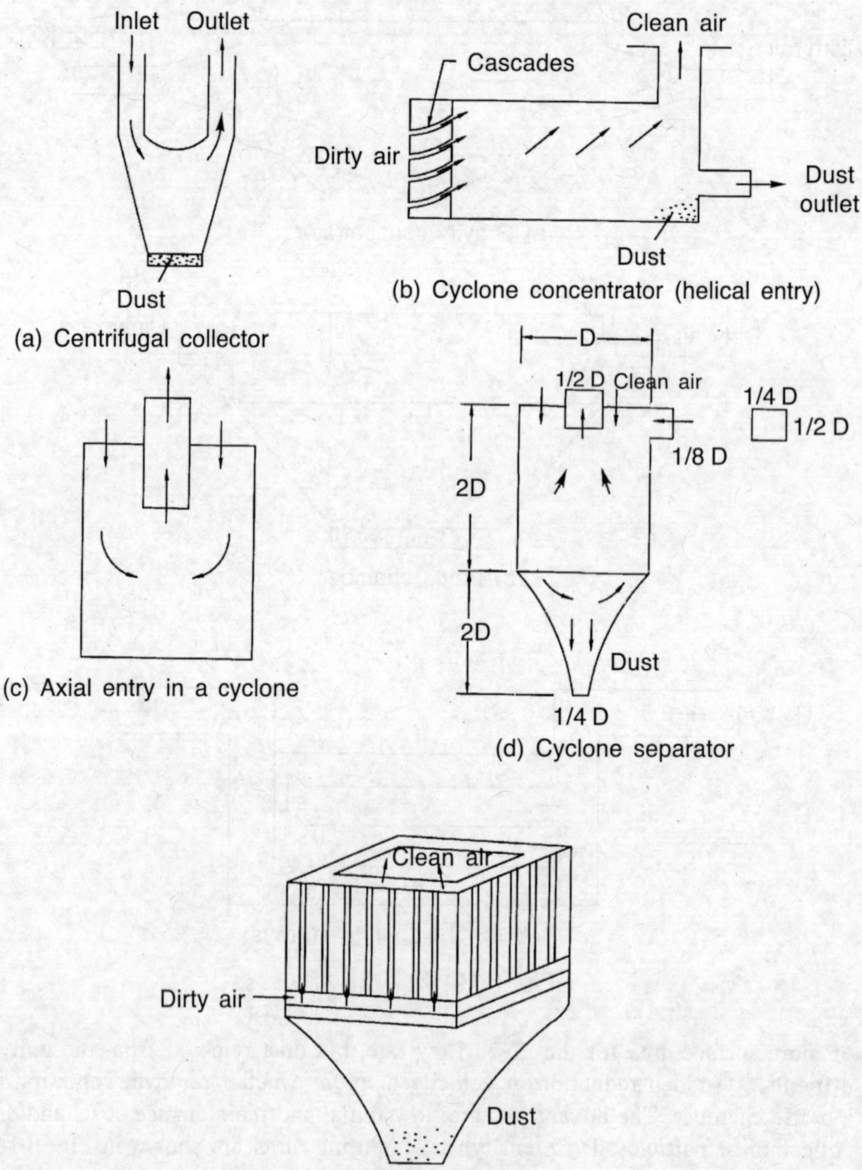

(e) Multiple cyclones

Fig. 6.15 Cyclone separators.

diameter of the cyclone is less. Standard shape and proportionate dimensions of a cyclone separator are shown in Fig. 6.15. Performance of a cyclone is affected by

- Temperature of influent gas
- Variations in flow velocity and volume
- Humidity conditions
- Proportioning the (dimensions) parts
- Shape, size and specific gravity and influent particle concentration

Bouncing and re-entertainment of particles, particle interactions, gas viscosity and density, may decrease the efficiency of a cyclone. Inside wall smoothness, formation of vortex, dust density and particle concentration help the collection efficiency. Wall surface may be affected by corrosion from reactive gases, sticky cake formation, or erosion due to attack from particles.

Cyclone separators may be used at high temperatures or pressures for wide range of particle sizes and concentrations. Multiple cyclones are large single compact units of smaller cyclones under parallel operation (simultaneously) to obtain higher efficiency of collection. Cyclones have an efficiency range from 60 to 90% but may not effectively remove particles of less than 20 micron size. There are no moving parts and therefore cyclones can be easily installed for handling large volumes of gas.

PROBLEM 5 Determine the diameter of the particle that is removed in a cyclone with 50% efficiency, given that

V (inlet velocity) = 16 m/s

$d \times b$ (inlet dimensions) = (1 × 0.5 m)

ρ_p = 1500 kg/m^3, μ = viscosity of air = 2×10^{-5} kg/m · s

Assume N (number of turns) = 5 or 6

Solution

$$d = \left(\frac{9\,\mu(b)}{2\,\pi\,NV\rho_p}\right)^{1/2} = \left(\frac{9 \times 2 \times 10^{-5} \times 0.5}{2 \times 3.14 \times 5 \times 16 \times 1500}\right)^{1/2} = 12\ \mu\text{m}$$

6.3.4 Wet Scrubbers

These are collecting devices where water captures dust particles and drain to sump. 0.1 to 20 micron size particles are removed in these units. Dry air movement and water sprays may be in the same, opposite (more optimum) or any direction. Spray towers, venturi scrubbers, and packed beds are shown in Fig. 6.16. Wet collection devices are preferred in foundries, mining operations and metallurgical

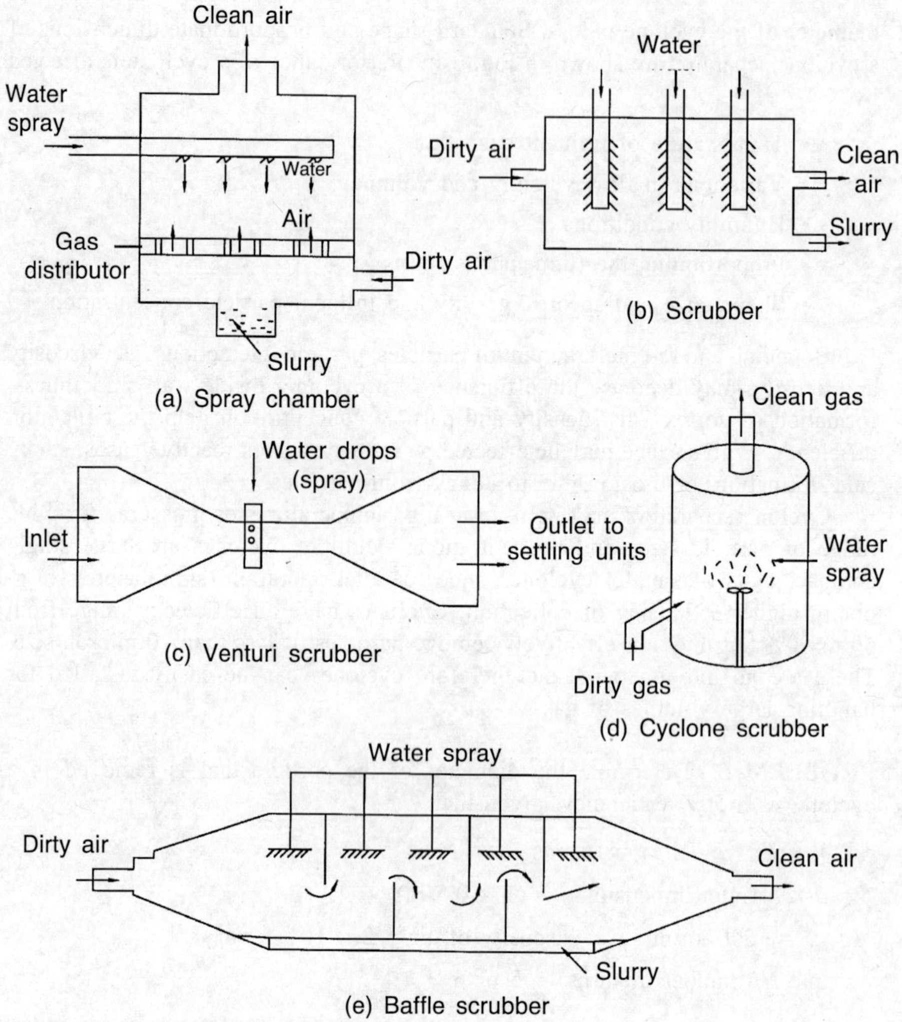

Fig. 6.16 Wet devices for particulate removal.

industry. In a simple spray chamber, water flows in the downward direction and traps the impurities. The chamber may be packed for improving the contact opportunity and collection efficiency. A baffled chamber may also be used for wet impingement. In a cyclone scrubber, polluted air is tangentially swirled around whereas water sprinkles into the chamber continuously. In a venturi scrubber, high velocities (60–100 m/s) at throat section atomise the liquid. Particles strike against moving droplets because of the velocity difference between the droplets and particles, droplets accelerate in throat section. The action is not complete in throat section but continues even afterwards till settling. Hence throat length is important for travel and collection of particles. This type of scrubbing device is very efficient for the removal of particulates greater than 0.5 microns, due to a better capture between high velocity collision particles and atomised water droplets.

Advantages of wet devices:

- Reliable with high collection efficiency
- Highly soluble gases like NH_3, HCl, HF are also removed
- Removes corrosive chemical dusts that may be difficult for removal in filters or electrostatic precipitators

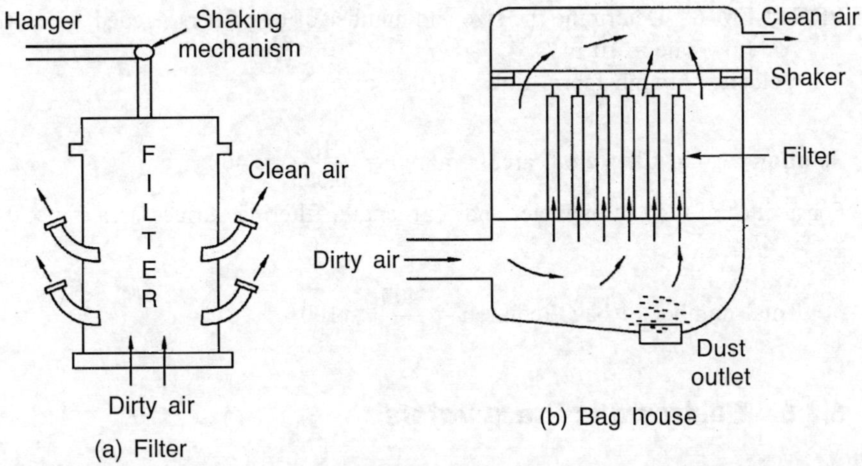

Fig. 6.17 Bag filter.

6.3.5 Bag Filters

These filters consist of porous structure material, fabric or fibrous matter which has the property, to retain the particulate matter as the dirty gas passes through the voids or interstitial spaces or layers of filter. Packed filters provide fairly long tortuous paths. In a bag filter (Fig. 6.17), filters are shown clamped around a sleeve at the bottom and cap at the top. Air enters from the bottom and flows out along its sides, leaving the dust to form a cake inside the filter.

Bags are usually of 15 cm diameter, and 3.0 m long (length to diameter ratio > 20) and maintain 0.5 to 1 m/minute flow velocities. Semi permeable wooven or felted material is used. The filter material depends upon the gas temperature and the nature of (acidity or alkalinity) gas as it has to resist abrasion and chemical actions. Common materials used are cotton (natural, fibre, cellulose), wool, nylon, orlon, teflon (fluorocarbon), fibre glass, and stainless steel. Filter efficiency depends upon the density and depth of filter packing and the ratio of gas flow volume to filter cloth (surface) area.

The efficiency is estimated to be 95–99% for removing fine particles (1 micron size). Some of the operational features of a bag filter are as given below.

1. Cleaning is done by rapping, shaking or sending air in the reverse direction

(with high velocity) at periodic intervals.

2. Fine particles penetrate through fabric and also cause rupture of cloth, which may be difficult to locate, repair or replace.
3. Reactive gases and chemical particles attack the filter media.
4. As the filter cannot withstand very high temperatures, suitable coatings may have to be applied over the filter surface.
5. Humidity should be controlled since hygroscopic dust creates problems.

PROBLEM 6 Determine the size and number of bag filters needed, given that
Air flow rate = 10 m^3/s,
Velocity through fabric filter = 0.025 m/s,

Solution Total filter cloth area required = $\frac{10}{0.025} = 400\ m^2$

For a filter of 0.25 m diameter and 5 m length, filtering surface area = $\pi \times 0.25 \times 5 = 4\ m^2$

$$\text{Total number of bags required} = \frac{400}{4} = 100$$

6.3.6 Electrostatic Precipitators

These are widely used in cement, power and metallurgical industries because of their capability to treat large gas flows, under high temperatures and corrosive atmospheres and remove submicron particles.

Effluent gas passes in a high electric field and gets ionised. Negatively charged particulates move towards positively charged collecting electrodes and attach to the surface. Two essential jobs, while using the precipitators, are:

- to set up an electric field between the electrodes for the formation of ions, and
- to charge the particles and the electrodes oppositely to create migration and attachment of particles.

The following steps are involved in the functioning of an electrostatic precipitator.

1. Impart electrostatic charge to particles
2. Create electric field in the flow region
3. Particles develop force of attraction
4. Particles migrate towards oppositely charged electrodes
5. Electrodes collect the impurities
6. Particulate dust is removed by shaking or rapping electrodes
7. The dry particulate matter is collected in the hopper below

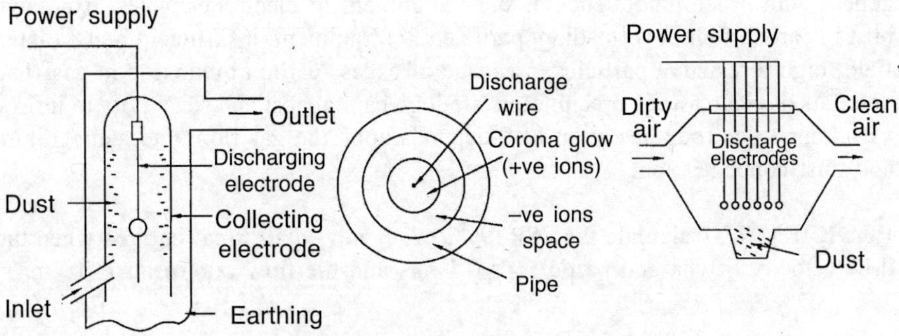

(a) Pipe type electrostatic precipitator

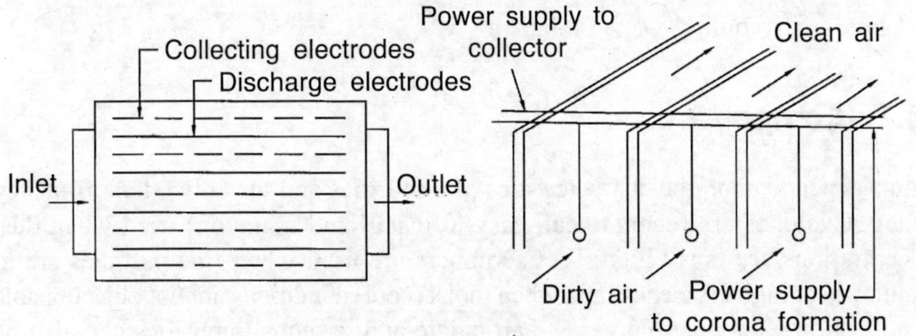

(b) Parallel plate type electrostatic precipitator

Fig. 6.18 Electrostatic precipitators.

Basically two types of precipitators are available—pipe type and plate type. In the pipe type, discharge electrode is a wire suspended (in a small pipe) with sufficient weights to maintain the position. Precipitation takes on the inner wall surface. In the case of plate type devices, collecting electrodes and discharging electrodes are placed in a parallel position as shown in Fig. 6.18. Corona forms between an active high voltage electrode such as a fine wire and a passive ground electrode, e.g. a pipe, or may be between two parallel or different plates. It is a mechanism of formation of ions in the gas. Dust particles carried through this space, instantaneously become highly charged and immediately migrate and attach to the collecting electrodes. Negative corona formation is more stable and efficient for collection. Corona characteristics depend upon:

- gas composition, temperature and pressure
- size, load and electrical resistivity of the particles
- Deposition on the electrodes
- electrode spacing and configuration

Electrostatic precipitators have a number of mechanical parts and electrical components and so have a number of operational and maintenance problems. But their efficiency of removal is almost 100% under design conditions, and removes particles greater and less than 1 micron size. A number of operational problems appear with precipitators such as poor alignment of electrodes/plates, excessive dust in corona field, overloading, particle re-entrainment, insufficient high voltage field, highly resistive particles, presence of excess carbon monoxide in gas. The demerits of electrostatic precipitators are high capital cost and sensitivity to inflow variations. Electrode design in ESP depends upon the gas flow rate, temperature and dust concentration.

PROBLEM 7 Calculate the ESP (99% efficiency) plate area required when the drift velocity of charged particles is 0.1 m/s and the flow rate of air is 10 m^3/s.

Solution Using the formula, $1 - 0.99 = e\left(-\frac{0.1}{10} A\right)$,

plate area required (A) = 460.5 m^2

6.3.7 Comparison

Choice depends upon the characteristics of impurities and the collection efficiency needed. Cyclones are (economical, easy to install and maintain) used when dust concentrations are fairly high. Wet scrubbers are used when fine particles are to be removed at high efficiency, and when moist (cool) conditions are not objectionable for collection. Combustible gases, particulate and gaseous impurities can also be removed. Fabric filters collect valuable material and are operated at low temperatures and volumes. Though efficiency is high, installation and maintenance costs are more. Electrostatic precipitators can handle large volumes of gas, at different temperatures and conditions. Efficiency is very high but requires constant attention and care.

6.4 CONTROL OF GASEOUS IMPURITIES

Absorption, adsorption, condensation and combustion are the main processes by which gaseous emissions are purified in the industry.

Selection of a suitable equipment depends on the considerations of

- Volume of air and concentration of pollutant
- Recovery of valuables
- Simultaneous removal of particulates
- Cost and efficiency considerations

6.4.1 Absorption

Effluents pass through packed beds (Fig. 6.19) or liquid absorbents so that pollutants are removed. Absorption or scrubbing is a diffusion process in which gas molecules are transferred into a liquid phase. When soluble gas contacts solvent liquid,

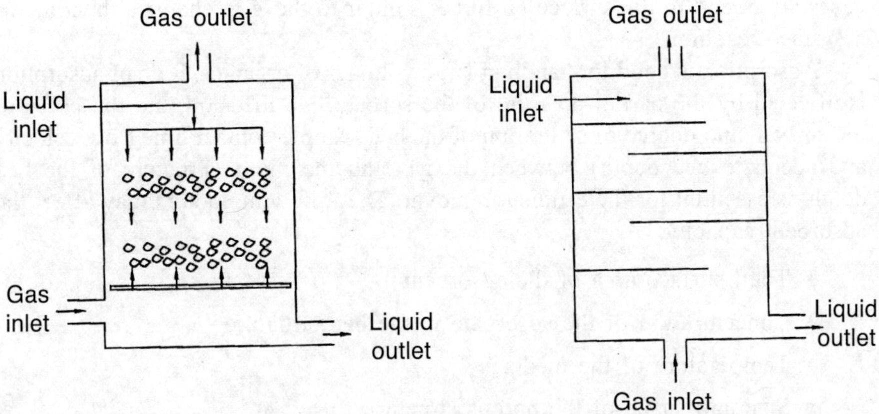

Fig. 6.19 Absorption towers.

mass transfer takes place due to difference in pressures of soluble gas in gas mixture and vapour pressure of solute gas in liquid film within the gas. Packed beds provide surface area and more intimate contact for effective absorption. Material may be recovered after absorption. Absorbents may be regenerated also, if needed, by stripping, oxidation, hydrolysis, solvent extraction, neutralisation and other methods.

The efficiency of absorption depends upon absorbing medium, contact opportunity provided in terms of time and surface area, and the chemical changes

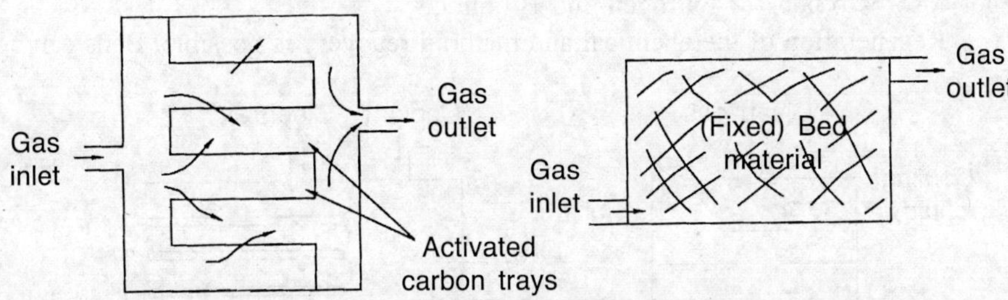

Fig. 6.20 Adsorbers.

involved. Water is a good solvent for NH_3, HCl, HF, Cl_2, NO_x, etc. SO_2 may be absorbed using lime or limestone, and iron oxide granules bed absorbs H_2S.

6.4.2 Adsorption

Undesirable gaseous effluents are removed in contact towers (Fig. 6.20), due to the adherence of molecules in a fluid state to the surface of a solid. Physical adsorption (or Vanderwaal's adsorption) involves a gas on the surface as a weak interaction in a surface phenomena which can be recovered back. In chemisorption, gases are bound to the surface on forces similar to those in chemical bonding as a surface reaction.

Adsorption is good for handling large volumes of organics. Rate of adsorption is affected by the partial pressure of the sorbate, its diffusion into the pores of the sorbant and degree of saturation of the bed. Ample contact time (low velocity and adequate bed depth) between the gas and the micro structure of the bed depth, is essential for the efficient removal. The following factors may affect the adsorbent capacity.

- Total surface area of the adsorbent
- Concentration of the adsorbate in the bed structure
- Temperature of the media
- Size and shape of the porous structure
- Adsorbate molecule characteristics
- Chemical activity of the adsorbent

(Freundlich, Langmuir and BET equations describe the adsorption phenomena).

Activated carbon (manufactured using coconut shells, bituminous coal or petroleum residues), siliceous compounds (silica gels, fullers and diatomaceous earth, synthetic zeolites), alumina, anhydrous $CaSO_4$, Ca or Mg silicates are used as adsorbents. Finely powdered activated carbon removes gases and bad odours too. It is best for dealing with organic vapours. Lime stone or alkalised ammonia is used for the removal of SO_2. Iron oxide bed removes benzene and H_2S. Silica gel adsorbs oxides of Nitrogen and Bromine.

Regeneration of the chemical and material recovery is possible. Beds can be

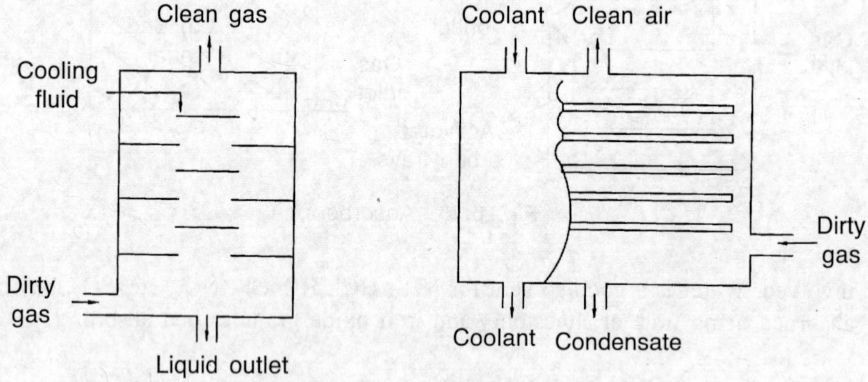

Fig. 6.21 Condensers.

desorbed (evacuation, stripping with inert gas, or displacement with more adsorbate material), if required. Fixed beds, travelling beds, and fluidised beds are used in the process of adsorption.

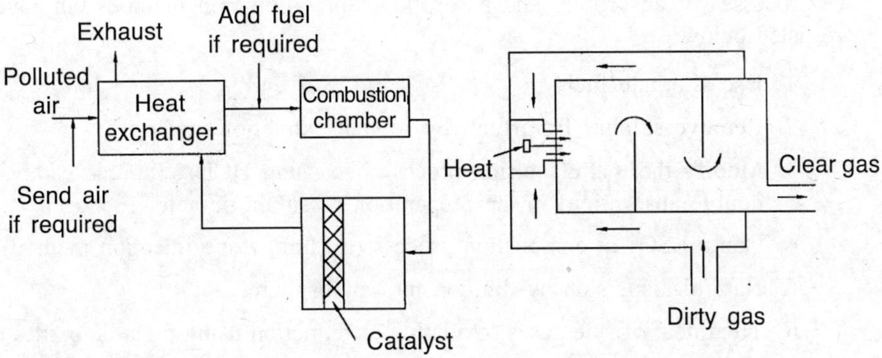

(a) Catalytic combustion (b) Thermal incinerator

Fig. 6.22 Incinerators.

6.4.3 Condensation

Condensation of a vapour from dirty air is possible either by increasing pressure or controlling the temperature. It is suitable for materials with low vapour pressures at moderately high temperatures, such as hydrocarbons. In a condenser (Fig. 6.21), coolant is circulated in the tubes. Air vapour stream circulates outside the tubes and vapour condenses on the cool surface and drops as liquid. This is collected at the bottom of a condenser.

6.4.4 Combustion

Incinerators are simple, safe and reliable. They are widely used for combustion of a variety of pollutant gases. Combustion of organics and gases, with the aid of auxiliary fuel and a catalyst enhances kinetics of combustion (Fig. 6.22(a)). CO and hydrocarbons are oxidised into carbon dioxide at higher temperatures (> 650°C) to achieve 100% results.

In the process of catalytic combustion, use of a catalyst speeds up the reaction at a desired temperature. Preheating is done in a thermal incinerator (Fig. 6.22(b)) when the pollutant air has a low heating value. In any case, addition of fuel or air, ensures complete combustion.

Combustion chamber (flame contact, temperature, residence time, gas phase, mixing), burner equipment (flow rate, hydrocarbons and temperature requirement), and refractories are to be properly planned and controlled for a cost effective system.

6.4.5 SO_2 Control

Various methods are available for the reduction of sulphur content from coal and crude oil used in industry. In general, gaseous emissions may be controlled by the processes of absorption and adsorption. Some common methods employed are listed below:

1. Use alternate fuels
2. Remove sulphur from fuel (by solvent extraction method)
3. Modify the manufacturing process like using DCDA (double catalysis double absorption) in the preparation of sulphuric acid.
4. Control SO_2 in combustion process (by lime stone injection method)
5. Natural dispersion by dilution into atmosphere.
6. Treatment of flue gases (oxidation or reduction using metallic oxides or absorption).
   (a) Recover sulphuric acid using vanadium pentaoxide as a catalyst at 400–500°C temperature.
   (b) SO_2 reacts with CuO in the formation of $CuSO_4$. Again CuO can be regenerated using H_2.

$$SO_2 + CuO + \frac{1}{2}O_2 \rightarrow CuSO_4$$

$$CuSO_4 + 2H_2 \rightarrow Cu + SO_2 + 2H_2O$$

$$Cu + \frac{1}{2}O_2 \rightarrow CuO$$

   (c) Use sodium aluminate in a fluidised bed.

$$SO_2 + Na_2O \cdot Al_2O_3 + \frac{1}{2}O_2 \rightarrow Na_2SO_4 + Al_2O_3$$

   (d) Scrubbing, using lime (lime stone) or magnesium oxide converts SO_2 into sulphates. CaO or $CaCO_3$ may be used in packed beds.

6.4.6 NO_x Control

Various physico-chemical processes have been developed for the recovery or removal of ammonia from effluents. Air or stream stripping is a standard practice. But other methods are also available, which are listed below.

1. Low excess air combustion
2. Decreasing air temperatures in combustion
3. Flue gas recirculation
4. Reduce NO_x by NH_3 injection
5. Injection of water or stream (stripping)
6. Modify configuration of the furnace burner
7. Liquid absorption using NaOH, MgOH, Lime or H_2SO_4

8. Adsorption by solids using activated carbon, silica gel, ion exchange resins and metal oxides like FeO
9. Non-catalytic homogeneous reduction by NH_3, when flue gas temperature is around 1100 K

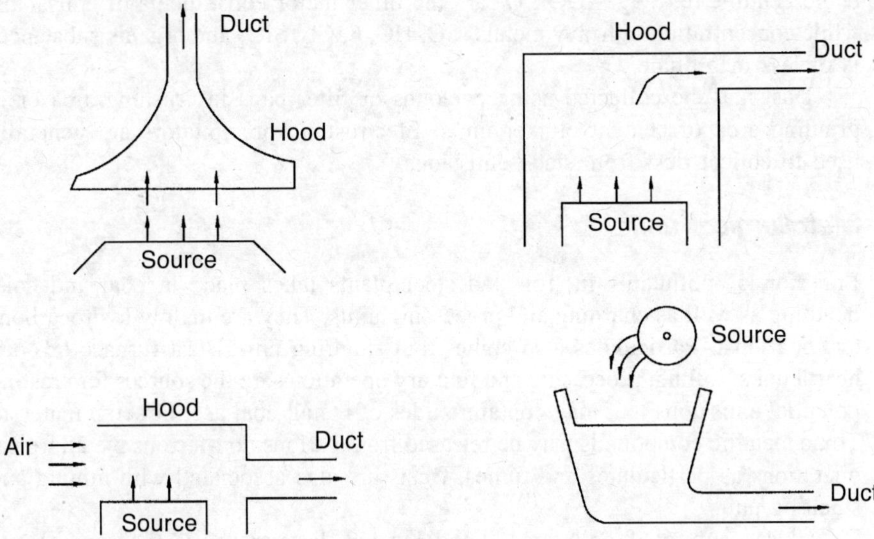

Fig. 6.23 Indoor ventilation.

$$NO + NH_3 + \frac{1}{4}O_2 \rightarrow N_2 + \frac{3}{2}H_2O$$

10. Selective catalytic reduction with NH_3 using V_2O_5 or TiO_2 at optimum temperature range of 600–700 K.

6.4.7 Indoor Ventilation

Natural ventilation is good in domestic areas for disposal of foul gases. Otherwise exhaust systems help. At industrial work places, pollutants are collected from the source through properly placed hoods and ducts which carry them away as shown in Fig. 6.23. Environmental sanitation measures including safety measures in occupational health systems are quite useful in preventing indoor air pollution.

6.5 SELECTED INDUSTRIES

Detailed studies on the raw materials and process procedures, source of emissions and pollutant nature and concentration, help in controlling various emissions from industries. Recycling and material recovery operations are also useful.

Some information on air pollution control in Industries is presented below:

Cement industry

Dust is generated at all stages of the plant (production and distribution). Concentration of air borne dust from limestone crushing, kiln firing, clinker zone, coal mills, packing and transportation units is around 20,000 parts per cubic centimetre. SO_2, NO_x, CO_2 are the other major constituents of emissions, while concentration of heavy metals, CO, HC, F, Cl_2, SiO_2 and organic substances is of less magnitude.

Dust may be collected using cyclones or filter bags in crushing area or in grinding area of cement clinker units. Electrostatic precipitators are generally used to collect dust from stack emissions.

Metallurgical units

Emission of pollutants in iron and steel plants takes place in coal and coke handling as well as charging and processing units. They are mainly hydrocarbons and coal dust. Particulates also come from sintering unit. Blast furnace or open hearth units, rolling, processing and foundry operations are the sources for creating potential emissions that may contain oxides of metal, coal ash, or scrap material. Toxic metallic compounds may be released from various non-ferrous metallurgical operations as particulates and fumes. Dusts are also associated with mining and smelting units.

Close capture of pollutants (hood located directly above the source) and proper aeration helps in disposal of harmful substances from foundries.

Petroleum products

Crude oil is distilled, separated, converted and treated in various ways to get finer petroleum products. Smoke, particulate matter, hydrocarbons, CO, SO_2, NO, NH_3, are the main pollutants from the emissions, depending upon the refining process.

Vapours may be removed from dirty gas through condensers, scrubbers or adsorbers. Cyclones may be used for the removal of particulate emissions.

Power plants

Combustion of fuel in power plants causes significant amounts of air pollutants. Fly ash, SO_x, NO_x are the primary pollutants entering into the atmosphere. Carbon monoxide and unburnt hydrocarbons are also emitted, and the coal units produces lot of dust.

Electrostatic precipitators are used to remove various particulate emissions.

SUMMARY

- Pollutants in the air cause injury to human, animal and plant life and damage to materials and property.
- Carbon monoxide, oxides of sulphur, oxides of nitrogen, hydrocarbons

and particulate matter are the major pollutants from transportation, fuel combustion and industrial activities.

- Various atmospheric conditions may lead to acid rains, global warming, photochemical smog, green house effect, ozone layer depletion in stratosphere and inversion conditions.
- Horizontal and vertical movements of air disperse the air pollutants and reduce the concentration due to dilution. Plume behaviour pattern depends upon the prevailing lapse rate in the atmosphere.
- Particulate matter of more than 40 microns size are collected in gravity type settling chambers.
- Due to centrifugal forces, particles are separated from gas flow in a cyclone, with reasonable efficiency.
- Wet collection devices have higher efficiency and gaseous impurities also are removed along with particulate matter.
- Bag filters and electrostatic precipitators are used in industrial operations to achieve nearly 100% removal of particulate matter.
- Absorption removes various gaseous impurities and adsorption is quite useful in removing organic impurities. Lime stone, activated carbon and silica compounds may be used.
- Combustion is a safe and reliable process for the removal of all gaseous effluents. If necessary, auxiliary fuel and catalysts may be added.

REVIEW QUESTIONS

1. List out the pollutants causing air pollution and their sources of origin.
2. Classify and outline the characteristics of particulate pollutants.
3. Explain the reasons for the occurrence of various air pollution disasters (episodes).
4. State the harmful effects of SO_2, CO, Pb and photochemical oxidants on human health, vegetation and materials.
5. Explain the relationship between lapse rates and atmospheric stability.
6. Explain typical plume behaviour patterns under different conditions of stability.
7. List the factors that are to be considered while selecting a particular device for removing particulate matter.
8. Explain the principles of working of a cyclone.
9. State the merits and demerits of bag filter over other devices.
10. With a neat sketch, explain the functioning of an electrostatic precipitator.
11. List out the absorbents and adsorbents used for removing various gaseous

pollutants from dirty air.

12. What are the requirements of a good incinerator for the combustion of air pollutants.
13. How are SO_2 emissions controlled?
14. Write short notes on
    (a) Secondary pollutants
    (b) Temperature inversion
    (c) Global warming
    (d) Topographical features on air movement
    (e) Automobile exhaust
    (f) Venturi scrubber
    (g) NO_x control
    (h) Plume rise

OBJECTIVE TYPE QUESTIONS

A. *Each of the following questions has four answers. Select the most appropriate one for each question.*

1. Depletion of ozone layer in the troposphere leads to
   (a) Radioactivity (b) Heating of earth
   (c) Ultraviolet radiations (d) All of these
2. Electrostatic precipitators remove
   (a) Sulphur dioxide (b) Nitrogen oxide
   (c) Particulate matter (d) None of them
3. A mixture of gas and fine particles is known as
   (a) Mist (b) Fog
   (c) Smog (d) Smoke
4. Global warming is due to
   (a) Ozone depletion (b) Photochemical smog
   (c) Green house effect (d) Acid rain
5. Sea breeze occurs during
   (a) Early morning hours (b) Evening hours
   (c) Day time (d) Night time
6. Atmosphere is stable if
   (a) ELR is more than normal ALR
   (b) ELR is more than the SALR
   (c) ELR is less than ALR
   (d) None of the above
7. An undesirable plume dispersion pattern is
   (a) Looping (b) Coning
   (c) Fumigation (d) Lofting

8. Corona is a mechanism of
   (a) Creating a charge on the particle
   (b) Migration of the particle to the electrode
   (c) Formation of ions in the gas
   (d) Developing force of attraction
9. H_2S may be absorbed by a bed of
   (a) Lime (b) Activated carbon
   (c) Iron oxide (d) Zinc oxide
10. Fly ash comes from
   (a) Cement plant (b) Refinery
   (c) Power plant (d) Paper industry
11. Which is the main cause of air pollution?
   (a) Processing industry (b) Forest fires
   (c) Transportation system (d) Waste burning
12. Plume rise decreases with
   (a) Stack gas velocity (b) Stack diameter
   (c) Wind speed (d) Stack gas temperature
13. An ideal mixture of air and fuel is the ratio of
   (a) 12:1 (b) 18:1
   (c) 15:1 (d) None

B. *Fill in the blanks with suitable words:*

1. The device which can be used to control both gaseous and particulate matter is _______.
2. 0.3 ppm of SO_2 is equivalent to _______ $\mu g/m^3$.
3. A decrease in the diameter of a cyclone collector _______ its efficiency (increases or decreases).
4. Mercaptans and sulphides are released from _______ industry.
5. Particles of _______ minimum size settle due to their size and weight.
6. When temperature increases with altitude, it is known as _______ rate.
7. Combustion is a process suitable for the removal of _______ from dirty air.

CHAPTER 7

Monitoring and Analysis

7.1 SAMPLING

Sampling means collecting a representative portion of water, waste water or air from an area to ascertain its quality and characteristics. Samples may be collected from rivers, wells, treatment plants, reservoirs and taps. Waste water samples may be taken from treatment plants (before and after treatment), disposal sites and in polluted rivers or soil. Air samples are generally taken from ambient air (surrounding human environment), stacks, air pollution control devices and inside factories.

7.1.1 Purpose

The main purpose of sampling is to

(1) ensure the quality of water and air as per the standards prescribed by various agencies,

(2) know the nature and degree of pollution from various sources,

(3) adopt and design specific purification plants,

(4) verify the efficiency of treatment plants,

(5) identify the chemicals and characteristics responsible for specific hazards,

(6) check the industries for compliance (as per the regulations) of emission standards, and,

(7) to prevent health hazards and disasters.

7.1.2 Collection of Samples

It is necessary to have preliminary information, such as topography of the area, climatic conditions, land use patterns, location of industries, sources of water and waste water, to identify the possible types of pollution and likely parameters for measuring the same. This is done through reconnaissance and pollution surveys.

When the nature and source of pollution are approximately known, one or two litres of random, instantaneous or 'grab' sample is taken at the time of active release of the pollutants from suitable locations. Samples may also be collected at regular intervals of time or continuously for a specific period of 8 or 24 hours in some cases, to evaluate the average concentrations at the same location over a period of time. They are known as composite or integrated samples. Continuous monitoring is very helpful in controlling pollution levels. The sequence of some essential steps to be taken while samples of water and air are collected are presented here.

1. *Water samples* are collected from rivers at different depths, at different reaches and at different times. While collecting sample from a running tap water, the midstream portion is to be taken.

The following points are to be remembered while taking samples.

(a) Sterilised glassware must be used. Bottles must be rinsed properly before collecting the sample. Care should be taken to avoid accidental contamination from any other source.

(b) Sometimes samples are to be tested immediately, otherwise there is a chance of significant changes taking place in the sample, due to biological or chemical reactions.

(c) Proper preservation of sample is a must, if the samples are analysed after sometime (within 24 hours). Keeping the sample under freezing temperatures (around 4°C) will preserve the conditions at the time of collection.

(d) Proper labelling is needed. Date, time, place, location, purpose, etc., must be marked clearly on the sample bottle (or bags). This information will be useful at the time of interpretation of results.

(e) Separate samples are to be collected for the analysis of chemical parameters and bacteriological examination.

(f) The actual quantity of sample required will depend upon the various tests to be conducted.

(g) The sample should truly represent the characteristics and conditions at site. It is known as a representative sample. Unless this is done, there is no meaning for the indication of results.

2. *Air samples* may be taken from various sources such as a stack (chimney), the ambient atmosphere, the pollution control devices and automobile exhaust pipes.

(a) High volume sampler (Fig. 7.1) consists of a glass fibre or synthetic filter of (about) 100 sq. cm exposure area, and pore-opening size of 3 microns. Suction pump draws air sample for a required period of time. Volume of sample of air passing through the filter is measured by rotameter. This sampler is used to collect particulate matter from the ambient air.

(b) Air is allowed to continuously pass through a sampling train (Fig. 7.2) consisting of various bottles (containers) having different absorbents or adsorbents for the study of various gaseous pollutants at a time. More parameters can be tested simultaneously.

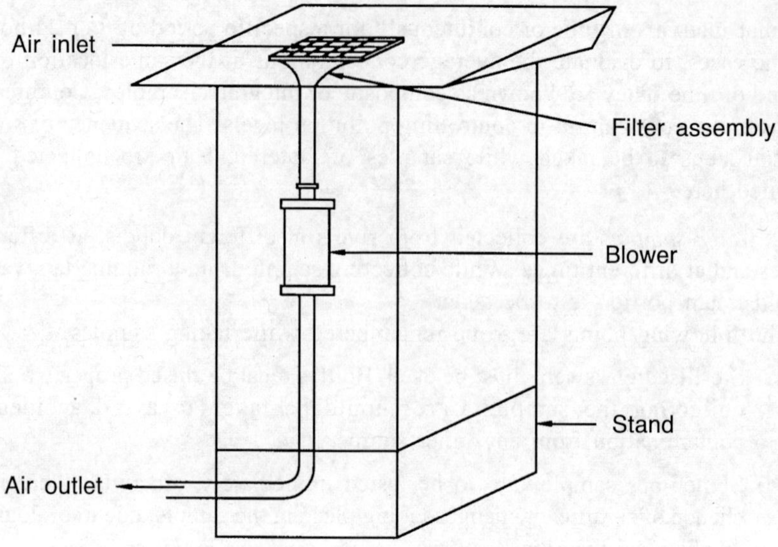

Fig. 7.1 High volume sampler.

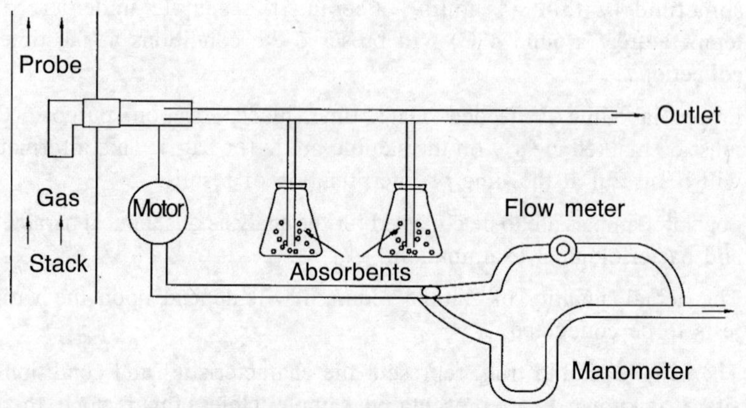

Fig. 7.2 Sampling unit.

(c) When polluted air moves through a stack, probes are taken into the stack for collecting samples. There may be some disturbance when probe interferes with the air movement inside. The velocity of air movement in the stack and in the probe should be same so that there is smooth flow. Such type of collection gives an isokinetic sample. Then the quality and characteristics of the sample truly represent the nature of inside air through the stack.

7.1.3 Interpretation of Results

The information collected preliminarily through pollution surveys should also be considered while interpreting the laboratory analysis reports.

Results must be reported preferably in the same units (mg/L or ppm) and manner as prescribed in the permissible quality standards of water or air. The following general precautions and procedures must be understood while making analysis.

1. Sometimes high concentrations should be diluted (in case of high BOD) to obtain the results in the desired range.
2. Sensitivity of the instrument and suitable detection procedure are essential to measure very low concentrations (metals in water).
3. Results obtained may be erroneous if interfering substances (like fluorides in dissolved oxygen test) are not properly eliminated.
4. Each method may have its own limiting conditions, such as pH, temperature, reaction time, etc. which should be followed strictly.
5. All the analytical procedures must be followed according to the instructional manuals supplied with the measuring instruments and standard methods prescribed by the national laboratories.

7.2 GENERAL METHODS OF ANALYSIS

Volumetric, gravimetric, colourimetric and instrumental procedures are available for analysis of samples in an environmental science laboratory.

7.2.1 Volumetric and Gravimetric Analyses

In a volumetric analysis, a fixed quantity of the sample is titrated against known concentration of a chemical solution, under suitable chemical reagents and colour indicator. When the chemical reaction is complete, the total of reactants and products must balance. As the volume and strength of the titrant and the sample volume are known, the unknown parameter concentration in the sample is obtained.

In a gravimetric method, solid matter is separated from water by sedimentation, evaporation and filtration techniques. The weight of solids per unit volume of a fluid is determined.

The above two procedures are simple and can easily be conducted in a chemistry laboratory.

7.2.2 Colour Comparison Method

Many pollutants are measured in the environmental science laboratory, by colour comparison technique.

Under a set of conditions, the intensity of colour developed is proportional to concentration of a particular chemical in solution, i.e. the more the concentration, greater is the intensity of colour. Relationship between light absorption (a parameter of intensity of colour) through a column length of the sample, at a particular wave length (depending upon the characteristics of absorbing substances) and concentration of the substance is linear according to Beer's law. So, the amount of monochromatic light passing through the sample solution is compared to that of a blank solution. When the transmission through the blank solution is 100%, the percentage transmission of an unknown solution depends upon the concentration of the sample. Hence suitable calibration curves are prepared (as shown in Fig. 7.3 for a typical case) for different concentrations of chemicals and their light absorbing intensities. An unknown sample is prepared under identical conditions and the concentration is obtained from the absorbance value as per the calibrated curve.

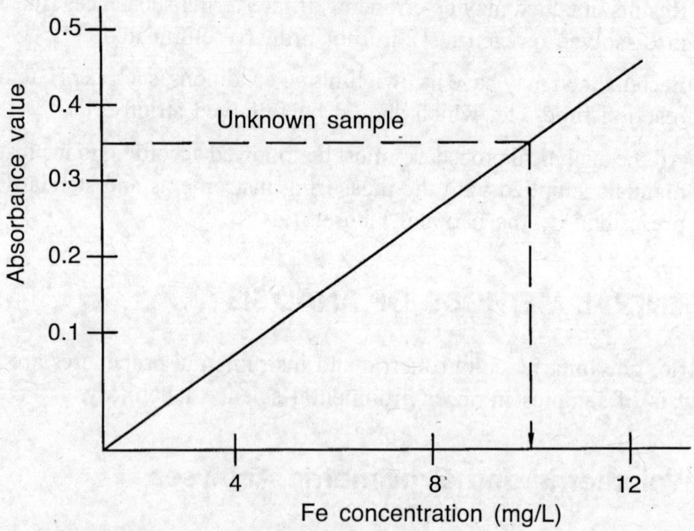

Fig. 7.3 Calibration curve

In a visual comparison, colours developed with unknown samples are matched with standard colours developed with known concentrations using a colour comparator (Fig. 7.4). Comparator is a small box containing two small tubes, one for the sample solution and the other for a blank, and a set of colour shades

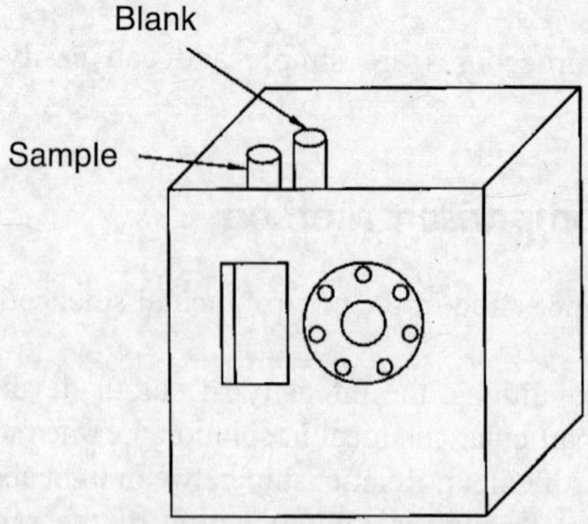

Fig. 7.4 Colour comparator.

corresponding to a given range of concentrations of a chemical in solution. The advantages of colour comparison method, is that a number of samples can be analysed in a short time. Instruments like spectrophotometers give the results more accurately.

7.2.3 Instrumentation

Various instruments are available for measuring directly a number of parameters in pollution analysis, such as alkalinity and chloride, fluoride and nitrogen contents. The three types of sophisticated instruments that are mainly used for measuring the concentrations of pollutants are discussed here.

1. In a spectrophotometer a lamp emits a broad radiation spectrum in the visible and ultraviolet range, and selected wave lengths are isolated and focussed. The intensity of light transmitted by the sample is measured by a photoelectric absorption having a detector and recording system.
2. In an atomic absorption spectrophotometer (AAS), atomic vapour is formed by spraying solutions into a flame so that solid solute particulates are converted into atoms which can exist in different energy states. By sending light of suitable wavelength, atoms passing through the atomic vapour from lower energy state to higher energy state are absorbed in that particular wavelength. Absorption characteristics of this vapour is a characteristic of that particular element. So, the desired element concentration is known from the absorption of monochromatic light using an AAS. This is highly useful in measuring the concentration of trace metals in water.
3. In gas chromatography (G.C.), (organic) sample is volatalised and injected along with the carrier gas stream (argon, helium, hydrogen) into G.C. column consisting of a polymeric sorbing material. While passing through this column, organic contents are separated as they possess separate retention times for the column material. Outgoing gaseous organic materials will be detected by a thermal conductivity or flame ionization detector. The detector responds according to the concentration of the material. Gas chromatography is more used for analysing organic pollutants in water or air.

7.3 ANALYSIS OF WATER AND WASTE WATER

In this section, analysis of most of the common parameters is presented so as to understand the basic procedures (also refer Section 3.1.3).

7.3.1 Physical Parameters

Colour

(i) 1 mg of platinum salt dissolved in 1 litre of distilled water produces 1 unit of colour on platinum cobalt scale. On that basis, prepare standard solutions of different units of colour. Compare the colour of given sample with the standard colours to obtain the intensity of colour.

(ii) Colour comparison can be done using standard discs of Hazen's apparatus.

(iii) Colour is due to salts in solution or in colloidal suspension. Colour is visible pollution and has an aesthetic value. Conventional water treatment removes apparent colour (due to the presence of suspended matter), otherwise removal is very expensive and difficult.

Turbidity

(i) Standard calibration curve is prepared using hydrazine and hexamine solution, having suitable dilutions.

(ii) Nephelometer reading with unknown water sample gives turbidity from a calibration graph, in *Nephelometer turbidity units* (N.T.U.).

(iii) The principle is that the amount of light scattered or absorbed is proportional to turbidity of the sample.

(iv) Turbidity may also be measured with a turbidity rod, *Jacksons turbidity meter* (Fig. 3.3) or other turbidimeters.

7.3.2 Chemical Examination

The sequence of steps to be followed to carry out the various categories of tests are given in the following sections.

Total Solids

(i) After known volume of the sample is filtered through standard filter paper, evaporate the filtrate and cool it. Obtain the weights of filterable and non-filterable solids in the above two steps.

(ii) Evaporate the water sample in a crucible to dryness in an oven at 103°C till the weight of solids remain to be constant so as to obtain total dissolved solids (mg/L).

(iii) Ignite the above residue at 600°C in a muffle furnace for about 30 minutes. Cool and weigh the solids. Loss of weight on ignition gives a measure of volatile solids.

Conductivity

(i) Conductivity meter gives a reading when the electrode is completely immersed in the water sample.

(ii) Multiply with the instrument cell constant to obtain specific conductance of the unknown water sample (micro mhos per cm) at the given temperature.

(iii) This gives an indirect measure of the dissolved salt content in the sample.

Acidity

(i) Add 2 or 3 drops of methyl orange indicator to 50 mL sample and titrate with 0.02 N NaOH solution till the colour changes to pale orange. Mineral acidity (mg/L as $CaCO_3$) is obtained from the quantity of titrant used.

(ii) Continue the titration after addition of 2 or 3 drops of phenolphthalein to a pink end point to obtain CO_2 acidity.

(iii) Acidity is the capacity to neutralise hydroxyl ions.

(iv) CO_2 and mineral acidity in water cause corrosion.

Alkalinity

(i) Add 2 or 3 drops of phenolphthalein indicator to a 50 mL sample and titrate against 0.02 N H_2SO_4 till the pink colour disappears. Phenolphthalein alkalinity (P as mg/L of $CaCO_3$) is obtained from the quantity of titrant used.

(ii) Continue titration after adding 2 or 3 drops of methyl orange till the orange colour turns to pink. Methyl orange alkalinity is known from the quantity of acid consumed in this titration.

(iii) Total alkalinity (T) is the sum of the above two. If the values of P and T are equal, then the total alkalinity is due to OH only and if $P = 0$, then total alkalinity is due to HCO_3 only.

pH

(i) pH meter is based on the principle that potential developed across a glass membrane due to the concentration gradient between a standard reference solution of Hydrogen ion and sample solution will be proportional to the pH value. Initially pH meter is adjusted with a known buffer solution (4 or 9). Later the electrode is immersed in the unknown sample and the reading directly gives the pH value in the proper range of 0–7 or 7–14.

(ii) Colour comparators are used (with a set of universal indicators and standard discs) for different pH ranges to match the developed colour of the unknown sample.

(iii) Various biological and chemical reactions depend on pH value of waters and waste waters.

(iv) pH is not a measure of total acidity or total alkalinity (Note the difference between degree of ionisation and neutralising value).

Chlorides

(i) Titrate the sample against 0.0156 N silver nitrate solution till the yellow colour turns red, using potassium chromate as indicator. Volume of the titrant used is a measure of chloride concentration in the sample taken.

(ii) Excess of chlorides indicates contamination from domestic and industrial wastes or salt water intrusion.

Total Hardness

(i) Add 1 mL of ammonia buffer solution to maintain pH 9.0 and few drops of Erichrome black T indicator to the water sample and titrate against standard 0.01M EDTA solution till the colour changes from wine red to blue. Total hardness (mg/L of $CaCO_3$) is obtained from the quantity of titrant used.

(ii) To obtain calcium hardness, titrate the sample against EDTA after adding 2 mL of NaOH and using muroxide indicator.

(iii) Hard water causes scale formation and consumes more soap in domestic use.

Chlorine

(i) Add a pinch of KI and 10 mL of acetic acid to the water sample and titrate against 0.025 N sodium thiosulphate using starch as indicator until the blue colour disappears. Available chlorine (mg/L) is known from the quantity of thiosulphate used.

(ii) Add orthotoludine reagent to the water sample and compare the colour intensity with the standards given, using a chloroscope (a colour comparator used for measuring the residual chlorine is known as chloroscope as shown in Fig. 7.4), to obtain the residual chlorine in water sample.

(iii) Chlorine demand is the amount of chlorine added to the sample, such that a residual chlorine of 0.2 mg/L results after 30 minutes of contact period.

Dissolved Oxygen (DO)

(i) Add 1–2 mL of $MnSO_4$ and 1–2 mL of alkaline potassium iodide—Azide solution to 300 mL sample taken in a BOD bottle. Mix thoroughly and dissolve the precipitate by adding 2 mL of conc. H_2SO_4.

(ii) Titrate 200 mL of the above solution with 0.025N $Na_2S_2O_3$ using starch indicator until the blue colour disappears.

DO in the sample (mg/L) is obtained from the quantity of $Na_2S_2O_3$ used in titration.

(iii) DO analysers may be used directly.

(iv) 8–10 mg/L of DO is present in drinking water. About 4.0 mg/L is required for the fish to survive. As DO is consumed by the organic matter, less DO indicates pollution.

Biochemical Oxygen Demand (BOD)

(i) Prepare proper dilutions of waste water samples with (1 L) distilled water to which 1 mL each of phosphate buffer, $MgSO_4$, Calcium and ferric chloride solutions are added. Add suitable biologically active seed.

(ii) Find out the immediate dissolved oxygen of the sample. Incubate few samples at 20°C for 24 hours (or 5 days).

(iii) Determine the DO again after 24 hours (or 5 days) of the incubated samples. The difference in the two tests gives the amount of oxygen utilised for biochemical activity of sample for the given time and temperature. Multiply with proper dilution factor.

(iv) Strength of the waste is indicated by the BOD.

Chemical Oxygen Demand (COD)

The following steps are to be followed to determine the chemical oxygen demand.

(i) Add 0.4 gm mercuric sulphate, 10 mL $K_2Cr_2O_7$ solution and 30 mL H_2SO_4 slowly to 20 mL of waste water sample in a flask.

Reflux the sample in a condenser for 2 hours. Cool and wash down the condensate with distilled water.

(ii) Dilute the mixture to about 140 mL, and titrate the excess dichromate with standard ferrous ammonium sulphate solution using ferroin indicator, till the colour changes to reddish brown end point of titration.

$$\text{COD (mg/L)} = \frac{\begin{array}{c}\text{Normality of ferrous}\\ \text{ammonium sulphate solution}\end{array} \times 8 \times 1000 \times \text{mL of titrant}}{\text{mL of sample}}$$

(iii) COD indicates the chemically oxidisable matter in waste waters. Biodegradability of wastes is indicated if the ratio of BOD to COD is greater than 0.6. TOC analysis measures total oxidisable carbonate matter in waste water.

Iron

(i) Draw a calibration curve (as shown in Sec. 7.2.2, Fig. 7.3) for the optical density with varying concentrations of iron using known samples (adding 2 mL conc. HCl and 5 mL of KCNS to develop colour) using a spectrophotometer.

(ii) For the unknown sample, find the optical density and obtain the corresponding concentration from the calibration curve.

Sulphates

(i) For known concentration of potassium sulphate solution, add 10 mL of NaCl or HCl, 20 mL of glycerol (alcohol) and make upto 100 mL using distilled water. Add 0.3 gm of barium chloride powder to obtain a turbidity. For different turbidities, prepare a calibration chart using a Nephelometer.

(ii) The absorption of the unknown suspension is measured and the sulphate content is known from the calibrated chart.

Phosphates

(i) Mix 2 mL of ammonium molybdate and conc. H_2SO_4 solution to 50 mL of water sample. Add few drops of 2.5% strength freshly prepared stannous chloride.

(ii) The blue colour developed is compared with standard colours to obtain the concentration of phosphate.

Fluorides

(i) Alizarin Red S forms a coloured compound with Zirconium vi, which will be bleached by fluoride ion as it forms stronger colourless complexes. The resulting bleaching of colour depends upon the concentration of fluoride in a water sample.

(ii) The colour intensity (similarly obtained) with an unknown sample is compared with standards to obtain the fluoride concentration.

Nitrogen

(i) Add 1 mL of $CuSO_4$ and 100 mL of conc. H_2SO_4 to 500 mL of sample. Digest the mixture till the solution becomes straw coloured. Cool and bring the volume to 250 mL. Add few drops of phenolphthalein and make the solution alkaline with NaOH. Distill the mixture and measure the volume. Compare the colour with that of standards (prepared with ammonium chloride solution + 2 mL of nesslers reagent made upto 100 mL using distilled water).

Total nitrogen (mg/L) =

$$\frac{\text{Volume distilled}}{\text{Volume nesslerised}} \times \frac{\text{mL of } NH_4Cl \text{ in standards} \times 0.1 \times 1000}{\text{mL of original sample distilled}}$$

(ii) To obtain ammonical nitrogen, distill the sample in kjeldahl flask (with 25 mL phosphate buffer solution). Add 50 mL of boric acid solution to the distillate and titrate with 0.02 NH_2SO_4 using methyl red indicator.

$$\begin{array}{c}\text{Ammonical} \\ \text{nitrogen (mg/L)}\end{array} = \frac{\begin{array}{c}\text{Difference of titrants} \\ \text{required in the above two cases}\end{array} \times 0.28 \times 1000}{\text{mL of sample (taken for distillation)}}$$

Settleable Solids

(i) Fill the Imhoff cone (Fig. 7.5) upto 1 litre mark with waste water sample and allow it to settle for 2 hours. Read the quantity of settled solids volume for every 30 minutes.

(ii) Collect the settled solids (after decanting the supernatant liquor) and obtain the weight of solids after drying in an oven. Express the amount of settleable solids as volume or weight per litre of sample.

(iii) This data is useful to understand the sludge characteristics such as specific gravity, moisture content, sludge volume index.

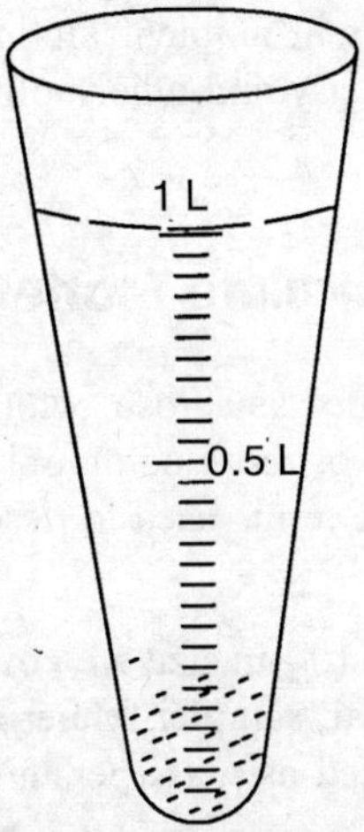

Fig. 7.5 Imhoff cone.

Sodium/Potassium

When a water sample contains Na or K, a flame photometer (salts are atomised) is used to measure the intensity of emitted light which is proportional to concentration of the element at a wave length of 590 nm for Na, [770 nm for K].

7.3.3 Bacteriological Examination

Three different procedures are available to carry out the bacteriological examination in a laboratory.

1. Take 10 mL of water samples in five test tubes each containing nutrient lactose broth. Keep Durham tube inverted in the test tube. Incubate all the test tubes at 37°C for 24 hours. Gas formation in the fermentation tubes indicates the presence of coliform bacteria (positive test) in the given water sample. Confirm the test after 24 hours with further incubation.

Record the number of positive findings of coliform group organisms and compute in terms of most probable number (MPN) (using a statistical chart) as a bacterial density in 100 mL of water.

2. Take Agar Agar (a nutrient medium for the growth of microorganisms) in a petridish with 10 mL of water sample. Keep the petridish, inverted in an incubator at 37°C for 24 hours. Observe colonies or groups of patches showing the presence of bacteria.

3. Filter 1 litre sample through a Sedgwick Rafter funnel. Take the concentrated material and mix the organisms carefully. Take 1 mL of the material on the counting cell and use a microscope for further examination.

7.4 AIR QUALITY ANALYSIS

Objective of ambient air quality monitoring is to collect basic data on quality and quantity of pollutants present in ambient air in relation to meteorological and other factors. Primary, secondary and other pollutants are analysed to determine the quality of air.

7.4.1 Suspended Particulate Matter

Particulates are generally termed as aerosol which may be liquid or solid. Liquid particles may be called mist, or more commonly, fog. Solid particles are called dusts, and smoke and soot are terms used to describe particles consisting mainly of unburnt carbon.

1. Particles of less than 10 μm size are collected through glass fibre filters, using a high volume air sampler. Material collected on the filters may be weighed and expressed as mass per unit volume of air drawn.
2. Particles of size exceeding 10 μm are measured by a dustfall jar of 15 cm diameter and 45 cm height. Samples are analysed once in a month and the dust fall rate expressed in tonnes per sq. km per month.
3. *Smoke* also contains suspended particles composed primarily of carbon, that result from incomplete combustion. Five ranges of colour from white to black at 20% intervals are shown in Fig. 7.6, Ringlemann's chart which is used for spotting smoke discharges from chimneys.

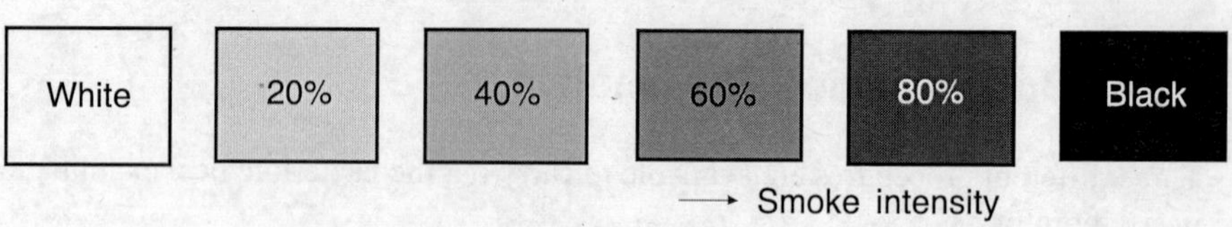

Fig. 7.4 Ringlemann's chart.

7.4.2 Gaseous Pollutants

Stack sampling is a method of collecting representative samples of pollutants in air or gases at the place of origin to determine total amount of pollutants discharged into atmosphere from a given source in a given stack. For collecting samples of flue gas it is necessary to (i) select suitable location for samples, (ii) determine traverse points required for and sampling temperature profile across the cross-section of the stack, (iii) determine accurately pressure, moisture and gas composition, (iv) measure rate of flow of flue gases, (v) carry out accurate isokinetic sampling of particulates, and (vi) measure accurately volume and weight of the sample collected.

Sampling point in the stack should be at a distance of 5–10 times of the diameter of stack from any obstruction or 3–5 times diameter upstream from similar disturbance.

Multi gas sampling kit is generally used in sampling different gases simultaneously. A small vacuum pump operates to draw atmospheric air through impingers when the gaseous pollutants react with absorbents. The chemical methods of analysis of the various gases are discussed in the following paras.

SO_2

(a) Reaction of SO_2 and lead oxide produces lead sulphate which is converted to barium sulphate for estimation gravimetrically or turbidimetrically. (sulphation rate)

(b) SO_2 is collected in hydrogen peroxide and H_2SO_4 solution and the increase in conductance of sulphuric acid solution is measured on a continuous basis by titration or using analytical instruments.

(c) SO_2 forms a complex salt with sodium tetra chloromercurate. Formaldehyde and pararosalinine dye are added to obtain a red purple colour (under controlled pH) the intensity of which can be measured at 560 nm wave length colorimetrically. (West-Gaeke method).

NO_2

(a) NO_2 forms sodium nitrite when absorbed through sodium hydroxide—sodium arsenite solution. Addition of sulfanalic acid and naphthyl ethylene diamine dihydro chloride results in red violet colour whose absorption at 550 nm is measured spectrophotometrically (Griess Saltzman technique).

(b) Reaction of NO and ozone (in excess) produces NO_2 in an excited state. When the latter decays to ground state, the luminiscence emitted is proportional to the original concentration of NO.

CO

(a) Absorbent silica gel impregnated with acid solutions of ammonium molybdate and palladous sulphate is allowed to be in contact with carbon monoxide. The colour intensity varies from yellow to blue green.

(b) CO is measured by the heat amount produced by its catalytic oxidation with a mixture of CuO and MnO_2. Heat generated is sensed by thermister. A vanadium catalyst can be used for the oxidation of HC and its analysis along with CO.

(c) Gas chromatography (with flame ionisation detection) analyses CO after conversion to hydrocarbons.

H_2S

Collect air sample in a wash bottle (containing 50 mL of 1% $CdSO_4.8H_2O$ and 50 mL of 0.01 M NaOH solution).

Mix 2 mL of 1 M $NaClO_4$. Add a mixture of

(i) 1 mL of 2% N diethyl P–Phenylene diamine in 50% H_2SO_4.

(ii) 1 mL of 10% ferric ammonium sulphate dissolved in M H_2SO_4.

Wait for 15 minutes and transfer the solution and make upto 25 mL. Measure the absorbance at 670 nm.

Calibration curve (prepared with standard stock solution of Na_2S) is used to obtain the unknown concentration.

Oxidants

(a) To obtain the total oxidant level of the sample, polluted air is collected through a solution of KI, producing I_3 which may be measured spectrophotometrically at 350 nm.

(b) Ozone reacts with rhodamine B on an activated silica gel surface and the emissions are measured at 590 nm region, in the chemiluminiscent method.

(c) A measure of oxidants indicates the level of Photochemical (secondary) reactions in the atmosphere.

Lead

(a) Collect sample on a 0.45 μm membrane filter (use I_2 crystals or carbon for trapping organic lead).

(b) Digest sample with HNO_3, H_2SO_4 and $HClO_4$ to solubilise lead. Dissolve I_2 crystals in acid KI and reduce with sulphite. Again lead is solubilised.

(c) Add NH_3 and extract with a chloroform solution of dithizone. The lead dithizone complex may be determined (at 510 nm) using a spectrophotometer.

NH_3

Ammonia develops colour with Nessler's reagent (K_2HgI_4 + KOH), and the intensity is matched with standard solutions to obtain the required concentration.

CO_2

Barium carbonate is precipitated when CO_2 is absorbed in a known excess volume of $Ba(OH)_2$ solution. The excess of barium hydroxide solution is titrated with standard hydrochloric acid. CO_2 concentration in air is known from the amount of unreacted $Ba(OH)_2$.

7.5 SOIL ANALYSIS

1. 1 kg of powdered, sieved and dried soil sample is prepared. Add sufficient water (about 5 times) and mix the contents. Wait for a day. Collect the supernatant water and use for chemical analysis.
2. Chemical parameters like pH, conductivity, alkalinity, chloride, nitrogen, phosphates, calcium, magnesium, sodium, potassium, iron and others are tested similar to procedures adopted in water analysis.
3. Bulk density, specific gravity, moisture content, porosity, permeability and other soil characteristics are determined in a soil testing laboratory.

SUMMARY

- The main purpose of examining water, waste water or ambient air is to assess their quality from the point of view of safety and environmental protection.
- Samples are taken from different sources, so as to truly represent all the characteristics at the time and place of collection.
- Volumetric analysis is suitable for testing acidity, alkalinity, chlorides, hardness and dissolved oxygen.
- Colour comparison procedure is adopted for measuring parameters like turbidity, pH, residual chlorine, iron, sulphates, fluorides.
- Various instruments are available for directly measuring turbidity, conductivity, pH, dissolved oxygen etc. They should be used as per the instruction manual.
- Atomic absorption spectrophotometer is useful in determining the concentration of trace metals in water.
- Gas chromatography helps in analysing organic compounds in fluids.
- High volume sampler is used to collect and estimate the particulate matter (also gaseous pollutants) in the ambient air.
- Chemicals react with plant, animal and aquatic life in a number of ways. Hence observations on such species also indicate the level of pollutants.

REVIEW QUESTIONS

1. Explain the terms
   (a) Grab sample (b) Isokinetic sample
   (c) MPN
2. Give neat sketches of
   (a) Sampling train (b) High volume sampler

3. Describe clearly the colour comparison technique for water sample analysis with a suitable illustrative example.
4. How is BOD determined in the laboratory?
5. What is the significance of the following tests?
   (a) pH (b) Settleable solids
   (c) Dissolved oxygen
6. State the working principles of a spectrophotometer.

OBJECTIVE TYPE QUESTIONS

Fill in the blanks with appropriate words:

1. The main purpose of water analysis is to ensure the quality as per __________.
2. BOD value is useful in treating __________.
3. Alkalinity is a measure of neutralisation with acid, whereas pH value determines __________.
4. In the determination of chlorides, __________ is added as an indicator, whereas orthotolidine reagent is used as an indicator in the measurement of __________.
5. Biodegradability of waste water is known from __________.
6. In the bacteriological examination of water, positive test indicates the presence of __________.
7. Gas chromatography is useful in the analysis of __________ as air pollutant.
8. Ringlemann's chart is needed in the measurement of __________.
9. Samples are taken from (name the parameter)
   (a) Cyclones to determine __________.
   (b) Activated sludge units to determine __________.

CHAPTER 8

Environmental Management

8.1 ENVIRONMENTAL IMPACT ASSESSMENT

8.1.1 Introduction

Environmental Impact Assessment (EIA) is a study to assess the damage, a project when implemented, can make on the environment. Projects (construction of a dam, a nuclear power plant, road flyover, etc.) are taken up only for the benefit of society at large. But when dams are constructed, reservoirs may submerge some villages. When a pharmaceutical plant is established, wastes generated may be toxic to the people living nearby. Liquid wastes from a tannery or other industry may pollute many useful agricultural lands or drinking water sources. There are always two sides of any project—the beneficial effects for the people and the harmful effects of pollution to the same human population. The benefits are always well-publicised. Nowadays, people are also aware of the evil effects. This impact or damage on the environment has to be analysed or assessed before taking up any new project. So, the main job of environmental impact assessment study is

1. to analyse all the evil effects on the total environment, on a rational basis, due to all project activities
2. to suggest suitable remedial waste management programmes, or modifications in the project activities to minimise pollution levels at source
3. to obtain necessary clearance and approval for the project proposal from the environmental considerations also
4. to select a suitable site for the project taking into consideration, environmental impact and cost/benefit analysis

A detailed study is required on the following aspects:

Status of the environment

This pertains to understanding the existing conditions around the proposed project location, before starting a new project or industry. Physical environment (water, land, air, etc.) and social environment (living conditions, historic or cultural monuments, moral and ethical values) should be studied from the considerations of living standards. Whatever may be the outcome (benefits) of a project, the present quality of life should not be destroyed.

Project activity

This requires understanding of the process details of an industry (or project), including the material (process reactions) requirements, and discharges from various units. Whenever a dam construction project is taken up, location of the Dam site, height of the dam or capacity of reservoir, though technical in nature, may lead to controversies. Selection of a particular route for a highway project may also be based on environmental considerations. In any case, there can be a viable alternative site for a project and that also should be analysed in detail.

Adverse effects

The disturbances (alterations), in the environmental set-up after the implementation of a project may produce some harmful effects. It is only a futuristic assessment. If a fertiliser industry is established, liquid effluents will be disposed into the nearby river affecting its water quality. If a dam is completed, there are villages likely to be submerged. Such ill effects on man, animal, plant life, resources, etc. are undesirable. For example, if the present level of fluorides that are likely to be discharged from a new industry could be assessed, it is always possible to understand the future fluoride levels in a nearby water source. From this, adverse effects on the local population, animals and plants/vegetation are estimated. But the fluoride levels can be reduced either by treating the wastes from the industry or treating water supplies. These project alternatives reduce the damage on the environment.

Pollution abatement

A number of control measures are available to minimise pollution at source. If treatment methodologies are incorporated in the project management programme, pollution of water, land, air can be reduced. Once the treatment technologies are included in the project plan, adverse effects or the damage function on the environment is assessed again with regard to the future set-up.

Damage function

This comprises two steps: to identify the cause and effect relationship, and to quantify the damage. That is, to identify the cause of damage, say fluorides, and next, whether the concentration of fluorides is 4.5 mg/L in water source. Then it is possible to estimate the ill effects of the pollutant on the population in a society. This damage is to be quantified in proper measure (units).

The harmful effects may be grouped or classified as

(a) Short-term or long-term effects. Exposure to high noise levels over a

long period of time leads to deafness, or drinking water containing more than 2 mg/L of fluorides over a period of time may lead to fluorosis. On the other hand, exposure to carbon monoxide gas in a sewer manhole even for a short duration or consuming toxic chemicals may be fatal.

(b) *Irreversible effects.* In the above cases, deafness, fluorosis, etc. are permanent effects on the human body.

(c) *Local effects.* If an industry discharges liquid wastes into a river or on land, the areas affected may be limited, affecting only the people living nearby. Ozone depletion in the atmosphere is a global phenomena.

(d) *Others.* The damages also may be categorized as significant or insignificant, primary or secondary.

All this is to distinguish between various impacts on the environment and evaluate the adverse impacts. Environmental impact assessment (EIA) finds a correlation between the various project activities and environmental attributes with an ultimate goal of protecting the environment. It is a study to estimate the likely future damage. It is a report on the industrial activities of environmental concern. It is essential for any project clearance from a competent authority.

8.1.2 Project Details

Projects vary in nature. It may be a construction of a flyover, a dam, a power plant or a distillery unit. Project details and activities are different in each case.

So, list out all the project activities step by step and identify the damage (if any) on the environment with each step or unit of activity. Also plan for alternate schemes and proposals. It is better to divide and subdivide various project activities further and further to the extent possible so that ultimate judgment on harmful effects on the environment is more accurate and specific. The following details are very important to estimate the side effects due to the implementation of a project.

Process details

Production flow charts indicate unit-wise material movement, chemical reactions, main products and by-products. Raw material needs or chemical requirements (quantity and quality) depend on the capacity of the plant. The nature of recovery or recycling in various units must also be studied.

Water quality and quantity

The amount of water and the quality of water to be used, and also the fuel and energy needs of one industry vary from the requirements of water and energy of another industry.

Waste discharges

Solid wastes (quantity and characteristics), liquid effluents (sources, quantities and characteristics) and gaseous emissions (sources and composition from various units of the industry) are to be ascertained. The collection, treatment and disposal facilities of the wastes and the management plans to mitigate pollution have to be incorporated in the project.

Other factors

Noise levels, odours, radioactive and thermal pollution (if any), occupational hazards, sanitation, safety measures are some more details which should be included in any industrial project.

It is also desirable to study a number of other (associated) project activities such as transportation of raw materials and the finished products, construction of buildings, and installation of machines.

The ultimate aim is to identify the damage on the environment due to various project activities. A project environment relationship is shown in Fig. 8.1. Shaded portion represents inside project area.

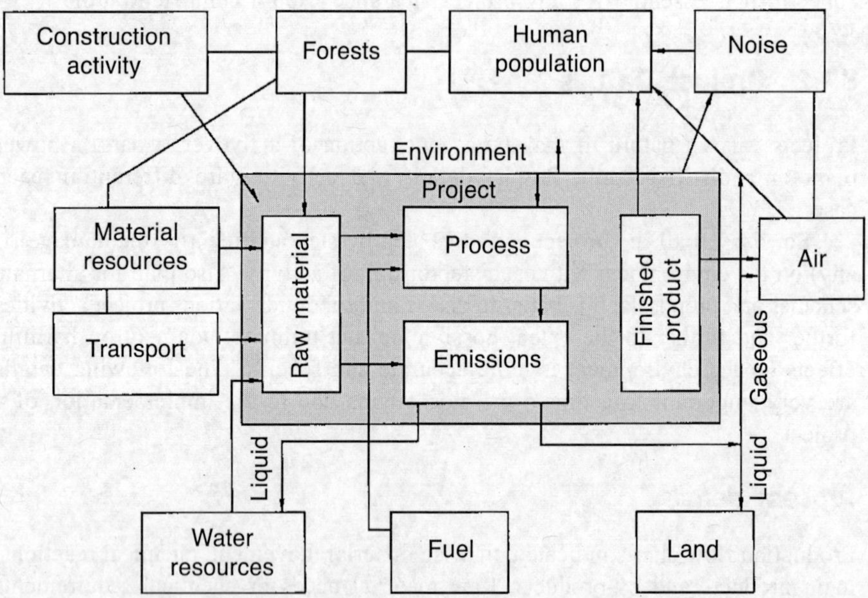

Fig. 8.1 Project-environment interaction.

8.1.3 Environmental Attributes

Environment can be broadly divided into two: a physical environment and a social environment. Physical environment consists of water, land, air, meteorological conditions, land use patterns, flora and fauna. Social environment consists of living conditions, community facilities, business potential and historic, cultural and human values. The number of attributes for consideration may be more and more depending on the purpose. These are termed as environmental parameters, or attributes which are of concern. However, if an attribute, say water is under consideration, the following aspects are to be considered under this parameter.

- Quality of water (physical, chemical and bacteriological)
- Sources, quantity, recharge potential
- Waste water quantity and characteristics at different point sources from the industry
- Waste water treatment systems available or proposed
- Future quality of water if the wastes (after proper treatment) are disposed onto water sources
- Water purification needed before use
- Harmful effects of the available water, on local population, animals and aquatic life (in future) in the nearby locality.

Further, *water quality* itself can be subdivided, on the basis of chemical parameters such as the presence of fluorides, chlorides, nitrates, sulphates, etc. Now, the ill effects due to each and every chemical, in excess of the permissible quantity, have to be estimated.

Thus, the total environmental attributes are to be divided and subdivided further so as to study and know how environmental quality is affected due to the various industrial activities.

It is also advisable to obtain information regarding the project site such as

1. topographic, geological and land use patterns
2. climate (temperature, rainfall and wind movements)
3. soil characteristics
4. wild life, forestry and related facts
5. existing living conditions in the surrounding areas, i.e. prevalence of diseases, social and cultural activities
6. self purification levels of land, river, air and other resources.

Finally, a chart is prepared incorporating project details and environmental attributes as shown in Fig. 8.2, for making an environmental impact assessment.

To fill up this chart, take one activity at a time; if it has any effect on water quality, write 'yes', and if it has no effect on air quality, write 'no'. If it has effect on some other environmental attributes, write 'yes' or 'no' accordingly. Then go to the next project activity and repeat the exercise. It would be better to evaluate the effects with markings such as, one or zero, significant or insignificant or, a 10–1 scale may be marked at the appropriate place to know whether that activity affects any environmental parameter, and if possible, to quantify on a suitable scale.

This information is quite useful only when all the project details, environmental factors and the effects are properly analysed and quantified.

Environmental attributes	Project Activities											
	Preliminaries			Construction activity		Processing units			Waste disposal systems			Others
	Site clearance	Temporary facilities	Transportation of Bldg. material	Construction work	Noise	Material sources	Unit 1	Unit 2	Solid	Liquid	Gaseous emissions	
1. Water environment												
Suspended solids												
Organics: (i) Degradable												
(ii) Non-degradable												
Inorganics: (i) Dissolved												
(ii) Heavy metals												
(iii) Pathogens												
2. Air environment												
CO												
NO_2												
SO_2												
Particular matter												
3. Land environment												
Soil characteristics												
Ground water contamination												
Land use patterns												
Agricultural operations												
Biological systems												
4. Social environment												
Antiquities												
Aesthetics												
Population changes												
Community infrastructure												
Employment												
Economic growth												

Note. Information on climate, fisheries, forests, mineral resources, topography, etc. also should be included in detail for a complete EIA study.

Fig. 8.2 Project environment relationship.

8.1.4 Impact Evaluation Methodologies

General requirements

Any system of EIA mainly consists of identifying the various effects on the environment due to project implementation at various stages and evaluating the possible damage in future. Two things should always be remembered.

- Environmental resources (considering also the self purification capacity) are limited.
- Social welfare (in all respects) must be the goal in implementing any project.

The following steps are followed in almost all EIA studies:

(a) As indicated in Sec. 8.1.2, visualise and isolate the various units, processes, stages of project (including alternate plans) and prepare a list of project activities.

(b) Identify the environmental parameters affected by the different project activities.

(c) Based on the cause and effect, estimate the extent and seriousness of the impact of project activities on environmental factors.

(d) Modify or alter the project proposals to minimise the damage on the environment.

(e) Predict the future changes in the environment taking into consideration the functions of various units of the project, and quantify the damage as accurately as possible.

(f) Use objective criteria to describe, interpret and evaluate the significance of cause and effect relationship.

(g) Clearly notice all risks, hazards and all proposals for compliance.

(h) Summarize and display the entire heterogeneous data systematically for further evaluation.

Methodologies

Some of the common strategies of evaluation are presented.

Ad hoc method. Experts (scientists, engineers and managers) review and identify the different project events and assess their possible effects on the existing environmental conditions in that particular locality. An overall expert view is reported. Their observations are not usually based on any detailed investigations. They may not also go into various alternatives and remedies available in a systematic and comprehensive method. But this report may be considered initially useful in the preliminary evaluation of project set-up from the environmental considerations.

Overlays approach. Several maps are prepared for the areas surrounding nearby localities of a project representing

(i) water and mineral resources

(ii) topographical and physical features

(iii) social and climatic conditions, and geographical status

(iv) existing industries.

When these separate transparent maps are properly superimposed, it is possible to know the places where pollution load is more. Thus, unpolluted areas can be located and visually identified for the location of sites for new industries. This method is also suitable for selecting the best possible route for a proposed highway project, from all the available alternative routes.

Figure 8.3 shows the result of superimposition for any three items of merit. Plots 1, 2, 5 and 12 are suitable for solid waste disposal from the considerations

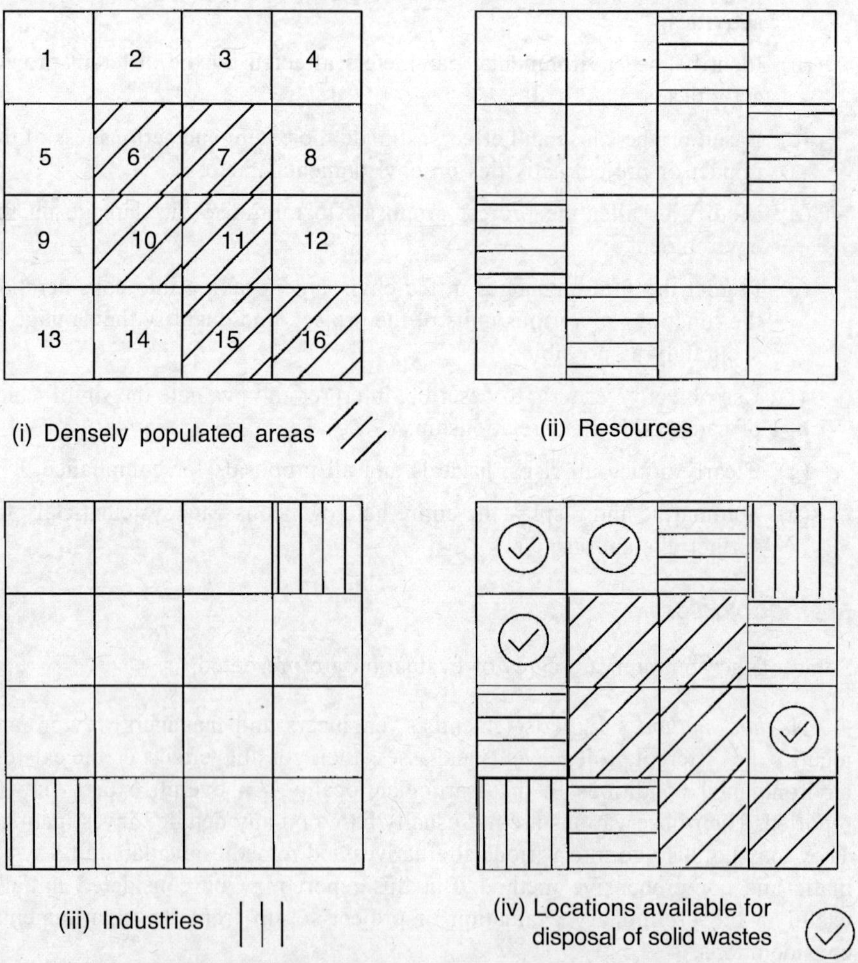

Fig. 8.3 Overlays methodologies for site selection.

of location of resources, industries and residential units. A shading system or colouring system is quite useful when various features are shown on a single map.

Networks method. From the item-wise list of project activities, direct and indirect effects caused on the environment due to each operation are estimated. Activities interrelated with others are identified and interactions between different activities are shown in Fig. 8.4. As networks are capable of presenting interrelated information, they are suitable for incorporating mitigation measures and management measures in the planning stages of a project.

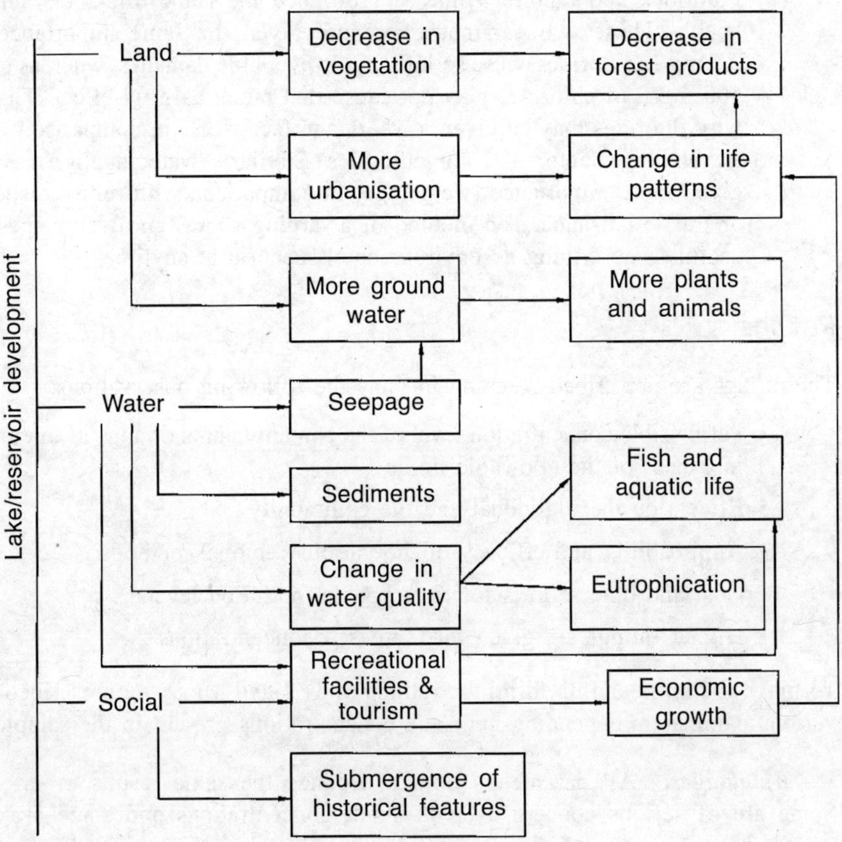

Fig. 8.4 Networks method.

Matrix method. Project activities and environmental attributes are presented in a tabular column as in Fig. 8.2. These matrices provide cause-effect relationships and impacts on environmental components. Any impact may be classified or described as adverse or beneficial, significant or insignificant, minor or major, to indicate the nature of impact. When the impact is marked as significant, adverse, major, yes-no, or one-zero, it will picturise the magnitude of the event or the relative differences between two impacts. Magnitude and importance of impacts may be evaluated on the basis of the following considerations.

(1) Fluorides are normally present in water in a concentration range upto 15 mg/L, whereas water containing less than 1.5 mg/L fluorides is safe for drinking purposes. Fluoride bearing water does have an impact on human health. Higher concentrations may create more problems in direct proportion to the level of concentration, or cause more and more troubles as in an exponential proportion. Accordingly, 5 mg/L may be rated with a rating index of 1, 10 mg/L as 2 or 4, 15 mg/L as 3 or 8 depending upon the expected proportionality factor. Similarly each parameter may be given a suitable score as per a rating scale.

(2) Fluorides and chlorides may not produce the same effect on human health. These two parameters are not given the same importance as 2.5 mg/L, fluorides cause serious and irreversible damages, whereas even 500 mg/L of chlorides may not cause that much harmful effect. That is why fluorides may be given a weightage factor of 5 in comparison with a weightage factor of 2 for chlorides. Similarly water quality may be given more importance (weightage) in comparison with refuse disposal or traffic nuisance. The method of awarding scores and weightages, is useful in describing an environmental scenario at anytime.

Ratings

The ratings are prescribed, keeping in view the following observations:

- Permissible concentration level of the contaminant and range of tolerance in excess of the allowable limits
- Effects on the individual and the community
- Applicability and efficacy of the various removal methods
- Cost and other efforts for reducing the pollution levels
- Higher ratings are given for increased concentrations

To make it simple and uniform, the ratings of 1, 2 and 4 may be prescribed for various parameters depending upon the concentrations present in the sample.

Weightages. All parameters do not produce the same results or effects. Some are of serious concern even in small concentrations and some are not removed in the conventional systems. Care must be taken to consider the relative importance of one parameter over the other.

So weightages of 1 and 2 may be given for each contaminant.

Procedure. Ratings are multiplied with weightages, for each parameter. Later, environmental quality index (EQI) can be obtained by

$$\text{EQI} = \frac{\Sigma \text{ Ratings} \times \text{Weightage factor}}{\Sigma \text{ Weightages}}$$

Points to remember

(i) All important quality characteristics are to be considered for calculating the index value.

(ii) Even if some parameters are not analysed, the quality index should be made available for interpretation with reasonable approximation.

(iii) When the concentration of one or few contaminants is very high, this fact should be significantly displayed, and probably, index presentation will not be of any value.

(iv) Computation of index must be an integral part of the analysis report.

(v) While assigning rating and weightages, the use of location for various purposes like agriculture, recreation, etc. may be kept in mind.

(vi) Bacteriological parameters need a special mention.

In this matrix method, the total environmental setting is described as shown in Fig. 8.5. The project is implemented only if the options are environment friendly.

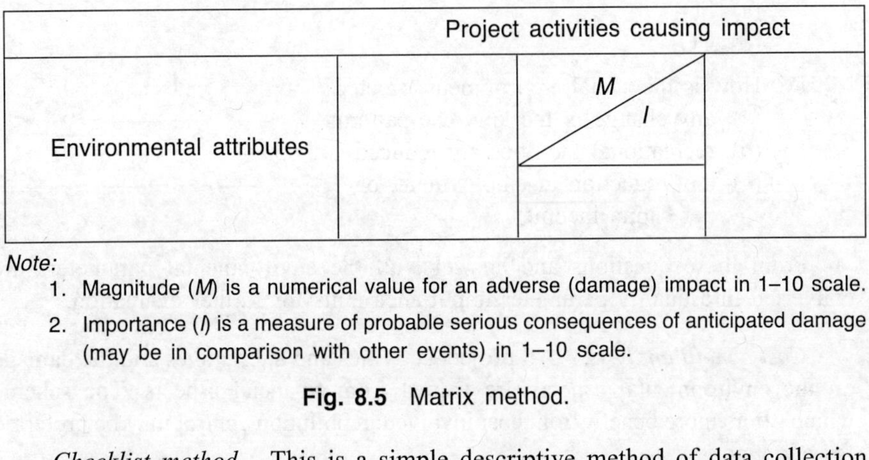

Note:

1. Magnitude (*M*) is a numerical value for an adverse (damage) impact in 1–10 scale.
2. Importance (*I*) is a measure of probable serious consequences of anticipated damage (may be in comparison with other events) in 1–10 scale.

Fig. 8.5 Matrix method.

Checklist method. This is a simple descriptive method of data collection in the form of a questionnaire. Hundreds of relevant questions are prepared to solicit answers. This information, when summarized in a systematic way, gives the overall view of the environment after project implementation. As each answer is related to the individual project activity and the related damage to the environment, the effects are given a numerical ranking on the basis of 1 to 10 (as explained above) or in the range of –5 to +5 (indicating harmful and beneficial effects).

Based upon the evaluation of each parameter individually and the relationship with every other parameter, the magnitude and significance of impacts are known.

For each and individual environmental factor, quality credits are obtained by multiplying the score value with weightage (parameter importance) factor. However, these ratings, scores and weightages are subjective. All adverse impacts cannot be brought significantly on to a numerical scale. Care should be taken to include secondary effects, interrelationships, project alternatives, pollution mitigation measures, etc. while framing the questions in the checklist. A few questions are mentioned below.

I. Does the proposed project affect the natural features of the surrounding environment?

(a) surface water quality	Yes/No
(b) soil erosion	Yes/No
(c) climate	Yes/No
(d) fisheries	Yes/No
(e) forests	Yes/No

II. Does the project cause environmental pollution (if yes, give more details)?
   (a) water
   (b) land
   (c) air
   (d) noise

III. Which parameter is of concern in the atmospheric environment, and at what concentration?
   (a) SO_2
   (b) CO_2

IV. How is the social environment affected?

(a) any change in the land use patterns	____	____
(b) recreational facilities are reduced	____	____
(c) any reduction in employment, or per capita income	____	____

From these questions and answers, all the environmental parameters are converted into quality parameter importance units for further evaluation.

Cost benefit analysis. All project activities and alternatives and the damage on the environmental aspects are valued from financial aspects. The scheme which offers more benefit from cost involved in pollution control may be preferred for adoption.

However, it is not good to consider only the financial aspects. It is not possible and reasonable to quantify human sufferings in monetary terms.

8.1.5 Limitations of Study

Comparison of various methods

(a) All environmental attributes may not be identified and quantified in the overlays method.

(b) Impacts and interactions are identified and displayed in a matrix method.

(c) In the network analysis, different impacts (social, human, aesthetic aspects) may not be properly distinguished.

(d) In check-list method, nature of impact may not be identified on the basis of inadequate questionnaire.

(e) In any system of environmental index evaluation, the following drawbacks are inherent.
   (i) assignment of weights and scores appears to be arbitrary in nature
   (ii) different items cannot be effectively transformed under one group
   (iii) aggregation process of all effects is irrelevant even if one factor is sufficiently hazardous in nature to stall the project.

Implementation of EIA

(a) Goals of any methodology are to be simple, flexible, representative, comprehensive and interpret the results accurately.

(b) Damage quantification, aggregation and organising large heterogeneous data is a super speciality job.

(c) Prediction of future damages on the environment are always subjected to limitations as there is no chance of accurate and actual environmental monitoring when the project is not initiated. Futuristic modelling studies are subjected to a number of uncertainties

(d) Mitigation methods may not be truly implemented to the extent provided in the EIA studies. Once the project is completed, pollution monitoring is not continuously done.

(e) Public should participate in the preparation of impact studies, as they are going to be affected due to the implementation of project, in that area. However, public opinion may be divided for different reasons.

8.1.6 Case Study—POWER PLANT

Four major steps in EIA

(a) Status report of environmental components like water, air, noise, flora, fauna, socio-economic studies, covering 10 km area around the proposed power plant location.

(b) Studies regarding the activities of the industrial units and release of various emissions from different sources.

(c) Proposed pollution control measures during various operations inside the plant and installation of treatment units.

(d) Evaluation of future damage on the environmental components due to industrial activity.

Pre-project environment

Data on the following parameters is collected near project site:

(a) Land—geology, hydrogeology, soil characteristics, porosity, permeability.

(b) Water—available surface and ground water quality and quantity.

(c) Air—wind direction, speed, humidity, rainfall, temperature, ambient air quality.

(d) Social—population distribution, literacy level, living standards, infrastructure facilities available, land use pattern, flora, fauna, noise levels, etc. (complete information is actually measured through sample studies at different locations in surrounding areas near the proposed project site).

Power plant operation

The following details should be made available for assessing the environmental impact of any power plant.

(a) working details of basic units

(b) water balance, quantity and characteristics of effluents, efficiency of treatment systems

(c) stack height, air pollution control equipment, flue gases quality and quantity, dispersion and dilution, impact on air quality

(d) fly ash disposal sites

(e) other factors like disposal of solid wastes, noise control measures, handling of hazardous chemicals. (This information is available from a detailed study of other existing plants of similar capacity and nature).

Check-list of plant operation phase

Environmental parameters Project activity	Land use	Water quality	Air quality	Noise	Solid waste	Other factors
Plant commissioning	X	X	X	X		
Raw materials handling		X	X			
Process reactions/units etc.			X		X	X

Note: X indicates possible impact.

Network analysis

Various impacts and interactions on air environment due to the proposed power plant are analysed through network analysis shown in Fig. 8.6. Similar studies on water, land and other environmental parameters present the complete picture of the total impact.

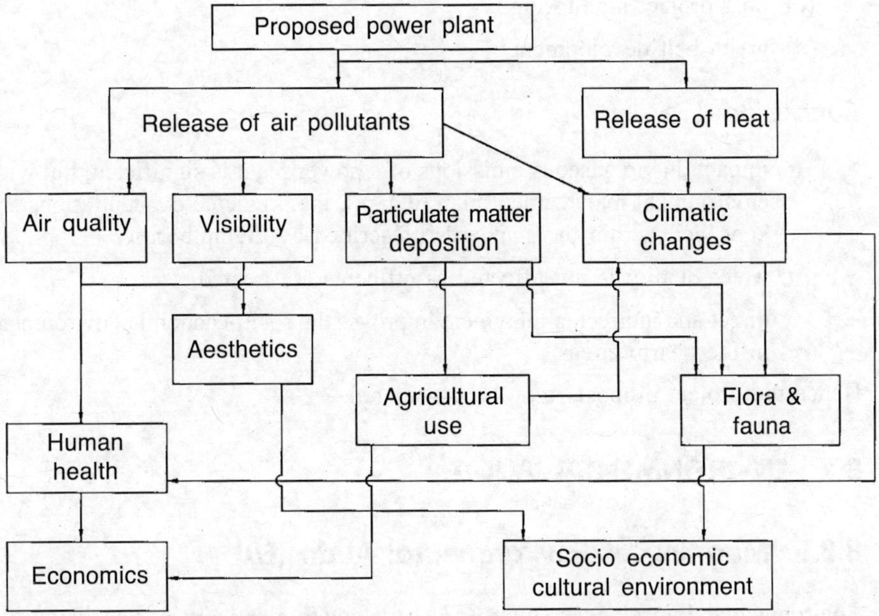

Fig. 8.6 Network analysis on air environment.

Summary evaluation

	Parameter importance units or weightage	Without project	With project		Changes	
			Without e.m.p.	With e.m.p.	Without e.m.p.	With e.m.p.
Environmental Pollution						
Air quality						
Water quality						
Land use						
Noise						
Socio-economic						
TOTAL	500	350	300	330	– 50	– 20
Ecosystem						
Aesthetics						
Human interests, etc.						
TOTAL	1000	700	650	725	– 50	+25

Environmental management plan (e.m.p.)

This includes measures proposed to be undertaken regarding

(a) storage of chemicals

(b) pollution control measures with respect to land, water, and air environment

(c) post project monitoring

(d) green belt development

Conclusion

- impact due to gaseous emissions of a power plant is significant, but with environment management plan (e.m.p.), the predicted concentrations are kept well within the standards prescribed by the authorities
- water quality is not affected as effluents are treated
- direct and indirect employment improves the socio-economic environment in the nearby areas

Hence the project proposal may be considered.

8.2 ENVIRONMENTAL AUDIT

8.2.1 Meaning of Environmental Audit (EA)

Environmental Impact Assessment (EIA) methodologies help in visualising the future consequences of an engineering project or a process industry. The harmful effects from the industry are contemplated very early in the planning stage, to search for alternative materials and processes. The impact assessment is a prerequisite before setting up an industry. When once the industry becomes operative, the organisation should take care of their own project to minimise pollution. The management has to systematically and continuously aim to regulate the processes within, to safeguard the outside environment. The procedure to review internally the management practices for the evaluation of pollution liability is known as *environmental audit*. It is a periodic, documented, and objective, self evaluation report of how the management systems and equipment are performing to minimise the pollution load on the environment. It is a job to assess the hidden risks, potential hazards and liabilities, to minimise a loss or leakage or accident and to effectively prevent and control pollution. Environmental audit helps in an early detection and rectification of a regulatory system by creating a database reporting system. Internal auditing system helps in safeguarding the health and safety of industrial workers. Finally, through environmental audit, a manager can implement the management plans for the recovery of useful materials and reduce the pollution potential.

'Environmental Audit' is a management tool to safeguard environment by (a) facilitating management control of environmental practices and (b) assisting compliance with company policies to meet regulatory requirements.

8.2.2 Audit Items

Total audit consists of a waste audit, an energy audit and a safety audit. The audit procedure includes the following evaluations.

1. Handling and storage of chemicals and process materials (including hazardous preparations).
2. Workers' safety during operation and maintenance.
3. Water (quality and quantity), raw material and fuel requirement.
4. Sources of waste generation, process discharges, effluent characteristics and flow quantity, ultimate disposal facilities, with reference to liquid and gaseous pollutants.
5. Collection and disposal of solid wastes.
6. Laboratory facilities for sampling and analysis of various pollution parameters.
7. Performance of waste treatment plants.
8. Neighbourhood survey to assess the impacts on land and water bodies.
9. Fire protection and safety procedures in emergencies.
10. Organisational set-up to deal with pollution control regulations.

The desirable conditions at any work place are:

- Safe working atmosphere
- Elimination of loss and leakage
- optimal material consumption
- Maximum output
- Conservation of natural resources
- Less pollution
- Recovery of useful materials
- Cost effectiveness in the total process.

8.2.3 Audit Procedure

1. The audit team consists of experts from technical, financial and management cadres.

   The four major functions of an audit team are:

   (a) Preparation of a checklist in the form of questionnaire
   (b) Reviewing the documentation and regulations
   (c) Interviewing the factory workers and
   (d) Physical inspection and verification of various items.

   It is also expected of the team to identify the problems and suggest solutions for a follow-up action.

   The job is accomplished in three stages.

   - pre-audit activities
   - activities at site
   - post audit data analysis

2. The following particulars are studied at pre-audit stage.
   (a) Location of industry, surrounding land use patterns, topographic, geographic, climatic and other conditions.
   (b) Raw materials used, details of process and unit operations, products and by-products produced, quality and quantity aspects.
   (c) Water requirements, solid, liquid and gaseous emissions including toxic and hazardous substances, waste generation sources, quantity, characteristics, and methods of treatment and disposal of waste.
   (d) Policy of the industrial unit with respect to environmental management and organisational set-up available in the industry.
3. The following on site activities are studied in different units.
   (a) Process unit operations, unit-wise material consumption, water balance, and input and output flow charts.
   (b) Pollution levels in various units, ground water, soil and air quality in the surrounding areas.
   (c) Solid wastes disposal and sanitation measures.
   (d) Evaluation of pollution control systems including waste minimisation, recovery and reuse of materials.
   (e) Opinion from workers, supervisors, managers in different units.
   (f) Management's point of view and actions to protect the environment.

   A draft is prepared considering all the above facts and findings, with specific realistic recommendations.
4. The following aspects are studied at post audit level:
   (a) raw material balance for each unit process (including reuse)
   (b) water balance (including recycling)
   (c) resource conservation
   (d) effluent quality with reference to emission standards
   (e) identification of waste minimising measures and problem wastes
   (f) efficiency of all waste purification systems.

   A final action plan is prepared as an audit report considering all the above aspects with specific implementation schedule. Environmental audit also helps the management to
   - adopt waste treatment systems
   - recover and reuse materials
   - implement management action plan (as an integral part of plant production) for controlling pollution
   - take steps in accordance with various environment related laws
   - plan sustainable development.

5. Audit proforma statement

   Proforma consists of factual statements, in different parts, well documented, including compliance programmes on waste minimisation.

   The requirements are outlined under different categories as:

   (a) name, address of location, industry category, production capacity, year of establishment

   (b) unitwise raw material and water consumption per month in processing and cooling operations

   (c) Effluent quantities discharged per day, concentration levels of pollutants and characteristics, steps for minimisation

   (d) Solid and hazardous wastes from process units, disposal and recovery systems

   (e) Financial requirements for pollution abatement

   (f) Conservation of natural resources

   (g) Additional measures to improve the quality of environment.

Environmental audit is Industry's own management plan to help in the evaluation of environmental protection measures periodically on a regular basis. It provides effective information for environment managers on the environmental quality and economics.

8.2.4 Safety Audit

This is an essential measure to prevent accidents and prepare the management to handle emergency situations. The following aspects are taken care of in safety audit.

1. Process safety includes measures taken to prevent corrosion, unwanted chemical reactions, high temperatures and evolution of toxic materials while using chemicals at every stage of the production process
2. Safety of structures and installations
3. Locational safety for materials from fire, explosion and electrical or mechanical installations
4. Fire protection devices
5. Measures to prevent occupational health hazards
6. Training of working personnel
7. Continuous supervision
8. Adherence to standard practices and statutory regulations.

8.3 POLLUTION CONTROL BOARD

8.3.1 Legal Aspects

One has an obligation to the society and has no right to do something that is injurious to others.

Rights, duties, liabilities and rules of human conduct and punishment for non-compliance of procedures, chartered by the State, form a part of Law. Indiscriminate disposal of solid, liquid and gaseous waste products alter the ecological balance and the dynamic equilibrium of the human habitat. Conservation of natural resources and ecosystems is absolutely essential for a sustainable development. No one is to be permitted to cause material damage on living and non-living components of atmospheric bodies. State has to protect and safeguard the quality of life.

Legal provisions emphasize certain procedures and rules regarding certain acts and also the penalties to be imposed for those not permitted by law. Any person polluting the environment deliberately, should be punished as permitted by the law. Water Act and Rules of 1974, the Air Act and Rules of 1981 and the Environmental Protection Acts of 1986, 87 are three major acts formulated by the Government of India. Workmen's Compensation Act, Factories Act, Forests & Wild Life Act, Motor Vehicles Act, Public Liability Insurance Acts, Insecticides Act, Hazardous Wastes Handling & Disposal Rules and various provisions of Sections of Indian Penal Code also provide relief and remedies for negligence and nuisance caused against environmental safety. All the courts have jurisdiction to implement the penal provisions and punish the guilty. These provisions will take care of all types of pollution and are uniformly applicable throughout the country. An industry may be closed if it is found to cause environmental pollution to the extent envisaged as per the provisions of the Act, or facilities such as water and power supply may be withdrawn.

8.3.2 Court Judgments

The following judgments are noteworthy.

1. Land and water sources are polluted beyond rectification around an industrial belt in Hyderabad. The court has given compensation to the affected people and directed the authorities to supply water to the villages regularly and recover costs from industry.
2. Polluting industries have been directed to shift their activities and units from Delhi within a framework of time.
3. Iron foundries near Tajmahal (Agra) are strictly directed to use pollution control methods or close them.

8.3.3 Functions of Pollution Control Board (PCB)

Pollution Control Boards are established at the State level, and the Central Board is at New Delhi.

The main functions of PCB and the objectives of the Act may be listed as follows:

1. to advise, plan and implement policies concerning matters of prevention and control of pollution
2. to compile and publish technical data, information manuals and code of practices
3. to organise, encourage, guide and assist in investigation, training and research programmes
4. to establish or to recognise the laboratories for analysing various parameters of pollution
5. to advise on selection of the location of an industry and to notify any area for industrial purpose
6. to issue environmental 'no objection' certificate to start an industry
7. to restrict the use of certain raw materials, to direct the use of approved appliances and to restrict emissions in excess of permissible limits
8. to exercise authority to collect samples for analysis and inspect the working conditions of pollution control equipment in any industry
9. to prescribe the effluent quality standards and to prohibit effluent discharges in rivers
10. to have safeguards against handling and disposal of hazardous chemicals
11. to prevent accidents and suggest emergency procedures
12. to frame rules and regulations needed to improve the quality of environment.

The Board consists of various members and officials who are representatives of the government, a Chairman and a Member Secretary. Persons from local bodies, such as fisheries, industries, trade unions, farmers, companies, etc. may become members of the Pollution Control Board. They may associate some persons also for getting expert advice.

8.4 ENVIRONMENTAL MANAGEMENT PLAN

8.4.1 Management Aspects

A Manager has to understand the following basic facts of sustainable development.

1. The basis of human existence is interdependency and maintaining biodiversity of an ecosystem.
2. Pollutants are likely to travel long distances through land, water and air.
3. Natural resources are always available in a finite quantity.
4. Renewable sources should always be developed and used alongwith non-renewable and non-conventional energy sources.

5. Future generations also have as much right to enjoy the environmental resources as the present population.
6. Environment has a limited self purification capacity.
7. Quality of life is improved if air, water and land environment is not contaminated with waste products.
8. Environmental protection is an integral part of all engineering projects and industry.
9. One has to protect the natural and cultural heritage of the country.
10. Pollution is a crime punishable under the Indian Penal Code.

For a safer environment, pollution of natural resources must be checked. This is achieved by the following measures.

- monitoring water, air, soil, etc. for assessing pollution levels in them at regular intervals
- minimising the generation of wastes and purifying the wastes before disposal
- recovering and recycling the materials from waste products, to make them reusable
- prohibiting the use of highly toxic chemicals
- strictly adhering to the safety practices.

If environmental and safety audits are carried out on a regular basis and implemented, it is possible to protect the environment. Incentives are offered to an industry for adopting pollution control measures and subsidies are provided for finished products.

8.4.2 Waste as a Resource

Some of the materials in waste can also be recycled or recovered for manufacture of new products or the same product.

Waste may be treated as a misplaced resource which is likely to assist in sustainable development. Environmental audit identifies the waste products and locates the sources of pollution in an industry. Based upon this information, environmental management plan is prepared, which is mainly aimed at achieving these goals:

- preventing or minimising the formation of waste at source
- recycling the less polluting waters
- recovering the valuable material
- converting the waste matter into marketable by-products
- utilising heat value to substitute energy consumption.

From a study and review of various successful industrial units and processes, a list of measures for reducing or minimising pollution is presented below:

- substitution of materials
- process changes or equipment modification
- recovery of materials
- reuse of materials, and
- possible financial gains.

1. Substitution of materials includes the following alternatives:

   use naphtha instead of benzol
   use non-mercurial compounds
   use desulpherised fuel
   use titanium dioxide instead of lead
   use carborundum instead of silica
   use bauxite flux instead of fluorspar
   use glass instead of plastics
   avoid sulphide ore in paper manufacture

2. Process changes or equipment modification comprises the following measures:
   - washing coal before pulverisation to reduce fly ash emissions
   - reduction in sulphur content in ores before smelting operations
   - using pelleted sintered orc
   - burning of coal by fluidised bed combustion to reduce sulphur dioxide emission
   - flow rate modification to reduce dust generation in rotary kilns
   - reducing NOx formation by low excess air combustion
   - wet grinding and brush painting to minimise particulate matter in air
   - using HCFC technology instead of CFC in polyurethane and polyethylene products.

Certain other precautions should necessarily be taken during the process changes,

- Build enclosures to limit the dispersion of pollutants in foundries, electroplating and other units
- Use personal protective devices such as respirators, gloves, shoes, goggles, aprons, etc. to protect industrial workers
- Maintain good house keeping, regular cleaning and operational schedules
- Adopt preliminary operations like separation of plastics and metals, shredding, dewatering, drying, etc., whenever required.

Useful building materials can be prepared from wastes such as:

- bricks from coal mine wastes, gypsum, fly ash

- paper boards from rice husk, bagasse
- wood substitute panels from cotton stalks
- concrete preparation using fly ash
- refinery sludge as a soil conditioner
- fly ash as fill materials for slushy lands.

3. The following materials can be recovered from various processing units:

Material recovered	*Units*
silver	photographic filming
zinc	rayon plant
Sulphur dioxide	copper smelting
Hydrogen, chlorine	silicon manufacturing units
Elemental sulphur	H_2S
Sulphur	power plants
Hydrocarbons and catalysts	petrochemical residues
Acids	resin manufacture
Copper and Lead	cables, and other units
Iron dust	steel works
Glycerine	soap factories
Solvents	chemical units

4. The following materials used in a particular process can be reused:
   - heat value from incineration of solid wastes, biogas from digestion units
   - water used in cooling and conditioning systems
   - non-condensable gases from petroleum refineries
   - cardboard from waste paper
   - fluorides from aluminium industry
   - pulp from white waters
   - plastics remoulded
   - recovery of ammonia

5. Financial returns can also be expected from various units by utilising the leftovers from the manufacturing processess:
   - animal feed from food, paper & distilleries
   - proteins from slaughter house
   - vegetable oil from sugar molasses
   - synthetic fibre from plastics
   - vermi composting manure from hatcheries, vegetable and fruit market, canteen wastes, etc.

- composting organic garbage with microbial stains from distillery spent wash
- briquettes from wood and agricultural residues with or without binders
- fuel pellets (from garbage) with good calorific value as a substitute for coal

(The entire list in this section (from 1 to 5) is provided from the data collected from a number of industries adopting these practices.)

From the foregoing discussions it is clear that any problem connected with waste collection, purification and safe disposal in an industry, has a managerial solution. Pollution control measures should therefore be adopted by environment managers to ensure the safe disposal of the wastes and at the same time bring some financial gain to the industry.

8.4.3 Objectives of ISO (International Organisation for Standardisation)

Environmental management systems have been detailed in ISO series

The main objectives for any industrial organisation are:

1. To have an environmental policy regarding environmental protection and sustainable development.
2. To initiate an environmental audit to assess their own performance.
3. To integrate the above two programmes in the comprehensive management activity plans.
4. To voluntarily declare how the organisation wishes to safeguard the interests of the surrounding environment in all respects.

However, adoption of ISO may not satisfy optimal environmental outcomes or occupational safety measures, unless the management truly performs pollution prevention and control activities in true spirit and without any excuses.

Environmental management system (EMS) recognises pollution management as a priority item. It assigns a commitment to protect the environment and the responsibility and accountabilities, considering the legal aspects associated with the company's activities, and products and services. The system also provides environmental planning throughout the life cycle of the product and process. It establishes management discipline for accomplishing the targeted performances and also evaluates the performance against policy targets besides emphasising adequate training to the employees allocating sufficient resources. There must be a process to review, monitor and audit the system on a continual basis. So EMS needs a policy formulation considering significant environmental impacts, to establish environmental protection measures and evaluate performances of all processes, devices and practices.

Requirements of EMS are as follows:

(a) A management policy towards
- nature and magnitude of environmental impacts

- commitment to prevent pollution
- complying with legislations and regulations
- framework and infrastructure needed for implementing the protection measures
- open policy document for employees and public.

(b) Planning job needs
- establish control procedures
- understand legal implications
- know financial and technological options, operational and business requirements
- have a specific time schedule for performing various operations.

(c) Implementation requires
- resources—skills, technological and financial
- appropriate education, training and experience and competency in the work activities
- documentation of all policies, impacts, protection measures, reviews, decisions and results
- operational control with regards to policy, objectives and targets
- to identify leakages, to prevent accidents, to adopt mitigation procedures, i.e. preparedness for all emergencies and response activities.

(d) Checks to be carried out:

(i) Monitor processes, control systems and environment on a regular basis and maintain a good record of analysis

(ii) Record all preventive measures, correction procedures, audit reports to demonstrate conformance to the ISO requirements, on a regular basis

(iii) Management should always put the following questions as ISO is based on self declaration:
- whether procedures envisaged in ISO are followed in the true sense.
- can the internal audit be relied upon?
- is EMS based on environmental audit?
- is EMS capable of achieving performance improvement and regulatory compliance?

The organisation (ISO) expects a positive attitude of the industry with regard to maintaining ecological balance and making environment friendly products. It reduces resource depletion. It minimises waste and pollution, and saves raw material. It promotes recovery and reuse of quality material, creates awareness among employees and community, and provides assurance, safety and savings. The policy of EMS is to protect the environment without fail. There is always a risk in not implementing EMS properly, as it causes legal liability, health damage, accidents, increased costs, wastage of raw materials, etc.

ISO is only the certification of quality management system of the organisation and is not product certification.

SUMMARY

- Some project (industry) activities are of environmental concern. But suitable measures can be incorporated to minimise the effects on physical and social environment, caused by various project activities (with suitable alternative proposals). EIA is a report submitted along with a project proposal to show that the project is environment friendly.
- Suppose one hundred activities of a project are listed and the affected environmental quality parameters are another hundred. EIA methodology analyses this 100 × 100 matrix, and estimates the damage function on each environmental attribute by various project activities.
- Environmental audit is an internal procedure to identify the sources of pollution and help the management in implementing the various control measures.
- Pollution Control Boards take all legal measures to protect the quality of life in the environment, as envisaged by the Water, Air and Environmental Protection Acts of the Government.
- Waste is also sometimes considered a resource from which valuable material can be recovered. These are generally cost effective and have high utility value.
- Environmental management plans (as envisaged in ISO series) are essentially management options to safeguard the interests of the management as well as the quality of the surrounding environment.

REVIEW QUESTIONS

1. Briefly outline the project activities of an industry (fertiliser plant or textile unit) that disturb the conditions in the sustainable environment.
2. List out the environmental factors that are affected due to the various activities of an irrigation project (construction of a dam).
3. What are the objectives of an impact evaluation methodology?
4. What are the major functions of an environmental audit?
5. Prepare a suitable checklist for data collection in the case of EIA for a flyover construction in a busy city road junction.
6. List out the advantages of environmental audit to an industry.
7. Describe the environmental evaluation system for converting environmental concerns into quality importance units that ultimately formulates an environmental index.
8. State the duties and responsibilities of a Pollution Control Board.
9. 'Waste is also a resource'. Give examples of recovery and use of wastes to establish the statement.

10. State the requirements of an environmental management system and the duties of an environmental manager in an Industry.
11. Write short notes on
    (a) Limitations of EIA study
    (b) Overlays approach of EIA
    (c) Polluter pays
    (d) Benefits of ISO

OBJECTIVE TYPE QUESTIONS

Each of the following questions has four answers; choose the most appropriate answer for each question.

1. Select the best environmental protection measure
   (a) EIA (b) EP Act
   (c) EA (d) ISO
2. In quantifying the damage, EIA considers
   (a) reversible and irreversible effects in the same category
   (b) a balance between gains and losses due to various activities
   (c) adverse effects on human beings as different from that on material resources
   (d) all pollution monitoring and controlling schemes
3. Select the wrong statement
   (a) EIA helps in choosing proper location of a project
   (b) EIA is only to obtain clearance from the government
   (c) EIA is a performance measure of an industry
   (d) EIA is a public document.
4. This item is not included in Environmental Audit
   (a) Storage of toxic chemicals
   (b) Pollution monitoring schemes
   (c) Scrutiny by the government agencies
   (d) Safety provisions for industrial workers
5. The duty of a Pollution Control Board is
   (a) to offer all consultancy services to the polluting industry
   (b) to continuously monitor the environment surrounding the industry
   (c) to prepare a code of practice for preventing pollution from an industry
   (d) to direct the industry to use a specific process for controlling pollution
6. Waste formation is prevented by
   (a) recycling less foul wastes
   (b) recovering valuable materials
   (c) substituting suitable new materials
   (d) isolating polluting units

7. Sustainable development is based on the fact that
   (a) Natural resources are plenty
   (b) Man made changes are only limited
   (c) Human habitat is an independent entity
   (d) None of the above

8. ISO methodology is a
   (a) management tool
   (b) voluntary declaration
   (c) documentation strategy
   (d) all the above

9. Environment is protected by
   (a) punishing the polluter
   (b) subsidising the needs of a farmer
   (c) monitoring pollutants regularly
   (d) awarding incentives for industries

Appendix A

TOXIC WASTES

Characteristics

Hazardous waste is the kind of waste which, because of its quantity, or its physical, chemical or infectious characteristics, can cause significant harm to human health or environment when improperly handled, treated, stored, transported and disposed. Hazardous wastes are generally toxic to any living being. Treatment methods are not readily available. Wastes from radioactive and explosive units are extremely hazardous and cause irreversible damage to environment.

Toxic wastes have four important characteristics:

- Ignitability (highly volatile and easily catches fire)
- Corrosivity (extreme pH values)
- Reactivity (violent, explosive nature)
- High toxicity (causes serious harm to living organisms depending upon their concentration level and exposure duration)

Harmful chemicals may be listed as cyanides, Hg, As, Tha, Cd, Pb, Cu, Zn, Cr, Se, Ni, Ba, phenols, asbestos, hydrocarbons, PCB. Further, toxic wastes may be grouped as, wastes from paints, dyes and other processing units, oils and emulsions, sludges carrying heavy metals, wastes from pesticides, acid or alkali slurries.

Special care has to be taken during handling, storage, transportation and disposal operations of these wastes. Some heavy metals viz., Cu, Pb, Cd, Hg, As, Cr, Ni even in very small quantities (less than 0.05 ppm) damage various parts of the human body such as liver, kidney, skin, respiratory tract and create neurological disorders. Some are even carcinogenic. On the other hand, many metals are also useful to life such as Iron (for blood), Ca and P (for bone building). Cu, Zn, Co, Cr in trace quantities play vital roles in the metabolic process of life. Various industries, agricultural operations and vehicular traffic are responsible for these heavy metals in water, air, soil, and food. Later they reach human body system. Sweets and beauty products may also contain toxic chemicals.

Some questions can therefore be raised:

(a) Is it possible to easily measure such low concentrations of toxic chemicals in food, water, air or soil?

(b) Is it possible to prove whether a particular chemical is responsible for a particular hazard?

(c) Are any alternative solutions available which are cost effective? (for avoiding use of these chemicals or eliminating them afterwards)

(d) Do legislative measures help in the prevention and control, of dumping of toxic chemicals into the environment?

(e) Is it possible to identify hazards and prevent them?

Probably the solutions are not that easily available.

Treatment Methods

It is always preferable to minimise toxic chemicals at the production stage by the substitution of alternate products, materials or processing technologies. Recovery and recycling or segregation may be used to reduce the strength (toxicity level) and volume of harmful substance. The waste may be detoxified or converted into a fixed (stable) state.

A number of options are available for treatment.

(i) *Physical methods:*
- (a) Air flotation (for removal of scum)
- (b) Evaporation (for concentration of dissolved salts)
- (c) Centrifugation (used for thickening of sludges)
- (d) Filtration (dewatering)
- (e) Absorption or adsorption (packed towers)
- (f) Reverse osmosis (higher concentrations reduced to lower levels)

(ii) *Chemical Methods:*
- (a) Coagulation and precipitation
- (b) Neutralisation (pH adjustment)
- (c) Oxidation and reduction
- (d) Ion exchange
- (e) Solidification (with binders)
- (f) Stabilisation
- (g) Disinfection (chlorination)
- (h) Fixation
- (i) Electrodialysis
- (j) Biodegradation
- (k) Thermal destruction

Final disposal alternatives are land disposal (after chemical recovery) or incineration (for organics) after providing the required treatment.

Land Disposal

Wastes are solidified, immobilised, compacted or dewatered to the extent possible.

They are contained (not allowed to flow or mix) and stored in controlled land fills. Care should be taken to minimise leachate flow and contamination of soil or ground water using impermeable clay liners as shown in Fig. A.1.

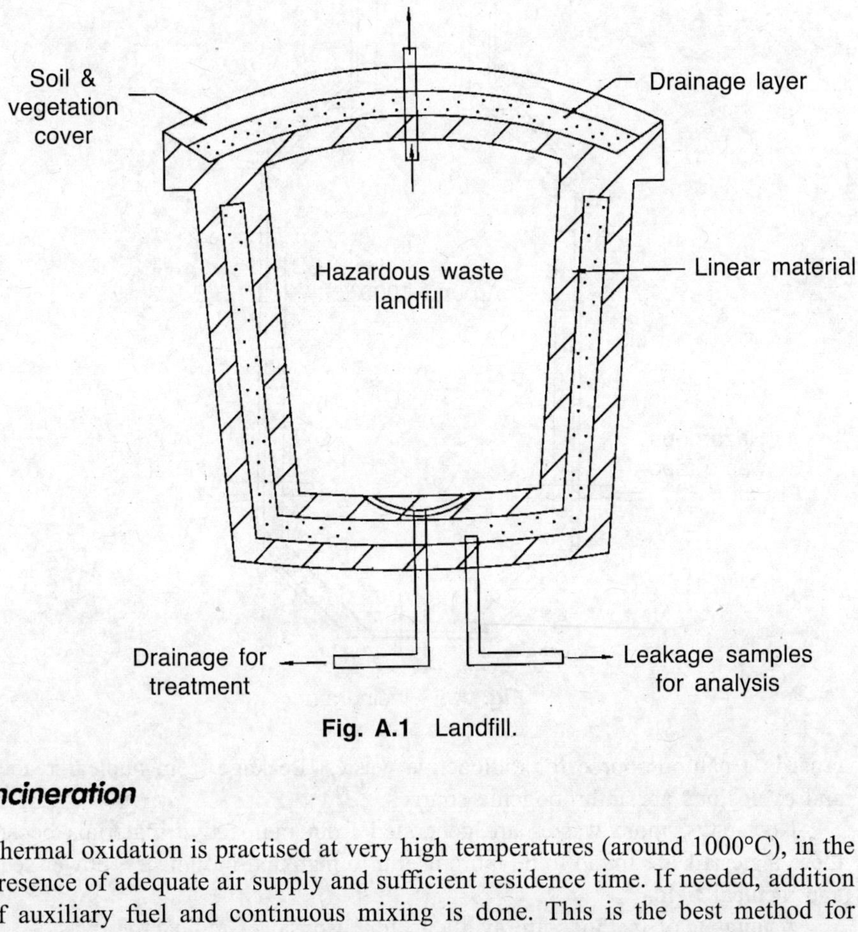

Fig. A.1 Landfill.

Incineration

Thermal oxidation is practised at very high temperatures (around 1000°C), in the presence of adequate air supply and sufficient residence time. If needed, addition of auxiliary fuel and continuous mixing is done. This is the best method for disposing organic wastes, pesticides, oils and biologically hazardous wastes (Fig. A.2).

Ocean disposal

In some cases, ocean disposal helps in dilution and dispersion of pollutants.

Radioactive Wastes

They reach air, water and soil and enter the food chain, causing serious damage to all living things. They cause many physiological changes in the subsequent generations (Hiroshima episode).

Weathering of atomic minerals, (thorium, uranium, radium), from the earth's

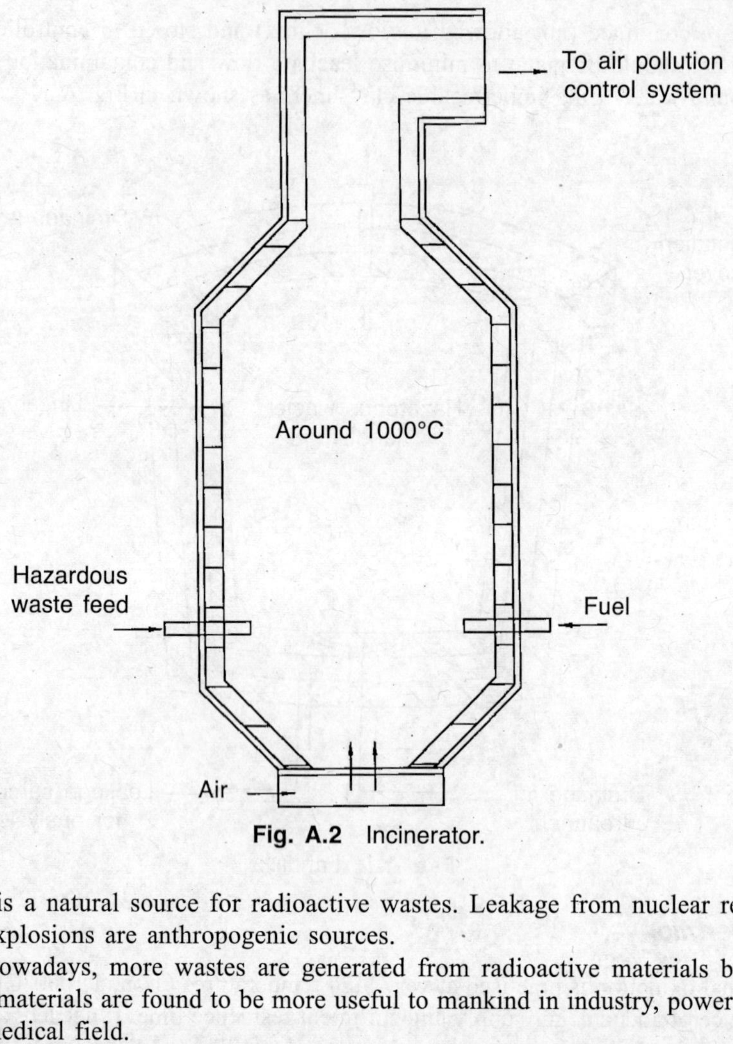

Fig. A.2 Incinerator.

crust is a natural source for radioactive wastes. Leakage from nuclear reactors and explosions are anthropogenic sources.

Nowadays, more wastes are generated from radioactive materials because these materials are found to be more useful to mankind in industry, power sector and medical field.

A number of methods are available for disposal of radioactive wastes.

1. Solidification and immobilisation; or evaporation and concentration to contain the waste so that they may be buried in underground stainless steel or special alloy containers or in deep sea waters (for a few centuries period).
2. Very low radioactive substances may be diluted and disposed in marine waters.
3. Medium active wastes may be imprisoned in concrete lined underground tanks so that wastes decay over a long period of time.

Appendix B

GEOGRAPHICAL INFORMATION SYSTEMS (GIS)

Maps

A map represents (to a scale) specific geographic locations and features like forests, lakes, hills, etc. Information is represented on the map in form of

- points (a town or oil well)
- lines (roads, railways, rivers)
- areas or polygons (boundaries of districts)
- contour lines (elevations of the ground) and
- colours (indicating forests, desert, mountain, lake)

Plane coordinate grids are used as a system of reference. This information, represented on a map, changes with time. So all the environmental features are to be interpolated periodically on the map.

Maps are prepared using

- The principles of surveying and aerial photography
- Satellite image processing and remote sensing techniques for data acquisition and processing and
- Cartography for graphical display of spatial information

Maps are created to different scales, different projections and in different colour combinations so that all the salient features are distinguished and assessed systematically.

What is GIS?

Geographic information system (GIS) is a methodology of displaying and representing features or attributes on the surface of earth at different time intervals. It is a computer based system for capture, storage, update, retrieval, analysis and display operations of spatial data. It integrates descriptive attributes and information from lists and reports (population, rainfall, etc.) with spatial data of measurement of geographical features. Finally, GIS analyses, transforms, derives and presents data needed for any specified purpose from spatial relationships.

GIS consists of the following components:

- Computer hardware (digitizer, scanner, peripherals, plotter)
- Central processing unit
- Sets of application software modules for data inputs and verification, storage and management, transformation and manipulation, and data output and presentation

Operational parameters involved in GIS activities are detailed as follows:

(a) Geographical data is stored in a number of formats including vector (lines, areas) representation, raster (an array of grid cells located by coordinates) representation. Data can be encoded using either manual or automatic techniques. Computer aided drafting systems capture and present maps in the form of electronic drawings.

Information (from field observations, existing maps, tables of attributes, photos, satellite imaging) is digitised (spatial features are converted into usable digital format) and stored as input data.

(b) Database management system organises the information (storage, manipulation and retrieval) into the required oriented structure. A database is a collection of information about things and their relationships to each other. It contains more than one attribute associated with its spatial location. Some details may be categorised under non-spatial information. The degree of relationship and the relational data model compatible with GIS are Info, Oracle, dBASE, and VB systems.

Updating, modifying and optimising are operations to edit the structure and contents of database. Sorting, indexing, classifying and extracting are the manipulations to integrate the database. Data processing also includes error detection and rectification. Querying generates useful information to establish patterns, trends and simulations needed for taking decisions. GIS provides for transformation of data to user specified solutions which is known as data analysis. Distortions can be removed, mapscales changed and new polygons generated to combine many data sets.

(c) Software functions create required types of output. Information from maps, figures and descriptive attributes are reported in the form of tables or displayed visually depending upon the requirements of precision and use.

Computer assisted mapping and design (CAD) aims at a software representation for input and output data including graphics. Automated mapping and facilities management (AM/FM) system provides a method of representing (storing, analysing and reporting data) maps and features in a graphic display or in data base form.

Applications of GIS

Separate source maps are prepared based on the thematic features like topography, natural resources, location of industries, population densities, rainfall, soil characteristics and the information is stored as layers. After analysis, desired output may be derived using the GIS relevant technique.

Problems like deforestation, depletion of water resources, status of ecosystem, etc. are analysed through GIS studies. If geographical information on the ecosystem before and after the construction of the dam is provided, changes in the environmental parameters can be easily understood. This data will be useful while making environmental impact assessment studies before any project of dam construction is taken up at any other location. Similarly, repetitive observations at a place indicate degradation of forests over a time period. Studies on land-use patterns, soil nature, water and air quality parameters indicate the pollution levels at a place in the course of time. However, field observations must be taken to supplement the requirements and confirm the trends.

GIS is used to identify suitable location to satisfy the given conditions for selection of a waste disposal site, construction of a dam or selection of an optimal route for a highway or railway project with less ecological disturbance.

Case Study

To identify locations for siting of industries based on environmental considerations. (refer p. 230, overlays approach).

Solution

1. Preparation of a base map of the zone, identifying various physical features regarding land use, drainage, constraints and other characteristics of the zone. Residential areas, water bodies, hills and contours, waste lands, agricultural lands, forests, mines, existing industries are to be clearly shown on the base maps.
2. Preparation of theme maps pertaining to water pollution (surface and ground), air pollution and land pollution. Physical constraints, social and legal restrictions should also be considered along with environmental considerations. Based upon certain quantification, high, medium and low categories of sensitivity (risk or level of pollution) are shown on each theme map.
3. Theme maps are overlaid one over the other and a combined map is drawn using GIS technique. These maps clearly depict the risks involved in siting the polluting industries in that zone under consideration.
4. Based upon the guidelines for location of a particular industry, possible sites are identified from the overlaid thematic maps after eliminating areas unsuitable for industrial activity.

Appendix C

DISASTER MANAGEMENT

Disasters

Earthquakes, landslides, cyclones, floods, fires, explosions, wars, riots, epidemics, accidents, leakage of hazardous chemicals, technological failures may be identified as disasters causing instantaneous damage on a large scale to the community. These uncertainties cannot be predicted with cent per cent accuracy. However, areas or locations, prone to each type of disaster, can be identified easily from the previous history of records. Coastal areas are more exposed to the fury of floods and cyclonic storms. Land slides, earthquakes are more likely to happen in mountain slopes and seismic zones of plateau regions. Flood waters stay for a number of days in agricultural fields and also submerge the whole human habitat.

During any devastation (a) human, animal, and plant life or the entire ecosystem is lost, (b) buildings and infrastructural facilities are severely damaged, (c) road, railway and other transportation system collapse completely, (d) power, water and telecommunication systems also fail, (e) agricultural produce and cattle feed are spoilt beyond use, (f) epidemic diseases like cholera, gastroenteritis, plague, etc. spread due to inadequate sanitation measures, and (g) finally, economic and social disturbances are created afterwards. It requires a lot of money, effort and time to restore normalcy. Rehabilitation is a very difficult task.

Preventive Measures

(a) land slides along the mountain slopes can be reduced by
- avoiding unauthorised constructions
- providing suitable drainage measures
- afforestation to the extent possible

(b) floods can be prevented by
- construction of suitable bunds and strengthening the same
- developing suitable vegetation along the banks
- diverting excess water through cross drainage works

(c) Building design needs a special attention. Foundations, roofs, walls, etc. should be properly taken care while selecting the materials and methods of construction such that the structure is cyclone/earthquake resistant.

Prior Preparation

Cyclone shelters

Construction of a special multipurpose cyclone resistant community shelter (or chain of shelters) for providing safe and secure accommodation is a necessity in the cyclone prone areas. These are constructed on an elevated ground with suitable provision of a community hall, kitchen, water power, communications, toilets, locker and other facilities. If required, nearby schools or public buildings may also be taken for providing additional accommodation. Cyclone shelters can also be used for community purposes (like adult education, vocational training, family welfare programmes) during normal periods. Bund protection as a flood control measure may be taken before monsoon season, with local public participation.

Information System

(i) Collect data regarding previous disasters, location, time of the event, atmospheric and other prevailing conditions at that time, extent of damage, steps taken during and after disaster, rehabilitation measures, failures, etc. This information provides a prior knowledge on the safe route of evacuation, time taken and difficulties faced while providing relief, problems involved in communication systems and quantum of relief needed.

(ii) Preparation of scientific database is very essential for forecasting a disaster. Then only it is possible to communicate the possibility of an event and issue warning signals for taking all precautionary measures.

(iii) Preparation of database on population, houses, cattle, agricultural and aquaculture farms, socio economic conditions of people, etc. help in evaluating the quantum of risk.

Activities before disaster

(i) training persons (home guards) with regard to the evacuation of affected persons and distribution of food and medicines

(ii) keeping fire, medical and ambulance services ready

(iii) arrangement of alternate shelter and communication works

(iv) mobilisation of medicines and relief material

(v) location of information centres

(vi) creation of public awareness and participation methodology

Relief Measures

(a) Immediately after the disaster, the first step is to cordon the area (if possible) and maintain law and order, with patience and presence of mind. The relief measures to be taken are:

(i) rescue and relief operations
(ii) restoration of power, water and communication systems
(iii) rushing of medical teams
(iv) supply of food, medicines and relief materials
(v) maintenance of sanitation
(vi) safeguarding agricultural products, live stock, and public properties
(vii) avoiding panic/rumour
(viii) keeping the morale of public high

(b) After some time has elapsed, the following steps are to be taken to restore normalcy to some extent.

(i) removal of road blocks
(ii) restoration of water, power and communication systems
(iii) repair of damaged houses
(iv) clearning debris to extricate the bodies
(v) restoration of the infrastructure
(vi) rehabilitation of destitutes
(vii) improvement of sanitary measures

Money, food and medicine may be available, but they should be diverted properly to the needy. All volunteers or rescue workers should act as a single team when any disaster strikes.

Bibliography

American Public Health Association (APHA) (1992), *Standard Methods of Water and Waste Water Analysis*, Washington, D.C.

Canter, Larry W. (1996), *Environmental Impact Assessment*, McGraw-Hill, New York.

Crawford, M. (1976), *Air Pollution Control Theory*, Tata McGraw-Hill, New Delhi.

Eckenfelder Jr. W.W. (1989), *Industrial Water Pollution Control*, McGraw-Hill, New York.

Euhlers V.M. and Steel, E.W. (1950), *Municipal and Rural Sanitation*, McGraw-Hill, New York.

Fair, G.M., Geyer, J.C. and Okun, D.A. (1971), *Water and Waste Water Engineering*, Vols. I and II, John Wiley & Sons, New York.

Mahajan, S.P. (1991), *Pollution Control in Process Industries*, Tata McGraw-Hill, New Delhi.

Masters, G.M. (2000), *Introduction to Environmental Engineering and Science*, Prentice-Hall of India, New Delhi.

Metcalf and Eddy Inc. (1991), *Waste Water Engineering: Treatment, Disposal and Reuse*, McGraw-Hill, New York.

Nemerow, V.L. (1963), *Theories and Practices of Industrial Waste Treatment*, Addison Wesley, MA.

Odum, E.P. (1983), *Basic Ecology*, Holt-Saunders' International ed., Japan.

Peavy, H.S., Rowe, D.R., and Tchobanoglous, G. (1985), *Environmental Engineering*, McGraw-Hill, New York.

Perkins, H.C. (1974), *Air Pollution*, McGraw-Hill, New York.

Rao, C.S. (1991), *Environmental Pollution Engineering*, New Age International (P) Ltd., New Delhi.

Rao, M.N. and Datta, A.K. (1987), *Waste Water Treatment*, Oxford IBH, New Delhi.

Rau Mohan G. and Guten David C. (1980), *Environmental Impact Analysis Handbook*, McGraw-Hill, New York.

Sharma, P.D. (1998), *Ecology and Environment*, Rastogi Publications, Meerut.

Sincero, Arcadio P. and Sincero, A. Gregoria (1999), *Environmental Engineering—A Design Approach*, Prentice-Hall of India, New Delhi.

Stern, A.C. (1968), *Air Pollution*, Academic Press, New York.

Wentz, Charles A. (1989), *Hazardous Waste Management*, McGraw Hill, New York.

Index